RADAR

for Small Craft

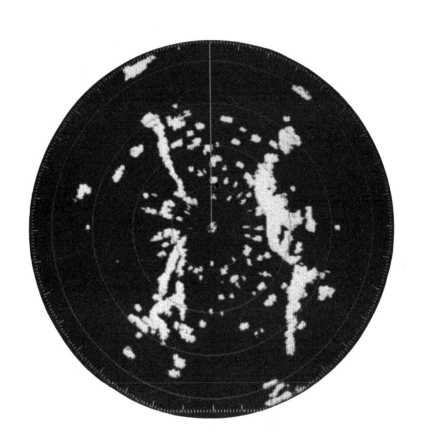

RADAR
for Small Craft

Tim Bartlett

© Copyright 1998, 1991 Fernhurst Books
Second edition published in 1998 by Fernhurst Books,
Duke's Path, High Street, Arundel, West Sussex, BN18 9AJ
Tel 01903 882277 Fax 01903 882715

ISBN 1 898660 48 4

Printed and bound in Great Britain.

Acknowledgements
The publishers would like to thank Raytheon Marine Ltd for
their assistance with this second edition and for providing
the cover photograph of their latest Pathfinder SL72 radar
system, Icom (UK) Ltd for the loan of a radar set for
photography, and *Motor Boat & Yachting Magazine* for the
use of their boat *Prospector*.
The portions of Admiralty charts in this book are Crown
Copyright, reproduced from Admiralty charts with the
permission of the Controller of Her Majesty's Stationery
Office. The portion of chart NZ 532 is reproduced by
permission of the Hydrographer RNZN.

Photographs
All photographs by John Woodward, with the exception of
the following:
Tim Davison: 43, 75, 76, 85
Motor Boat & Yachting: 33 (all), 48, 87 (left)

Edited and designed by John Woodward
Artwork by Pan Tek, Maidstone
Cover design by Simon Balley
Composition by Central Southern Typesetters, Hove
Printed by Hillman Printers, Frome

Contents

1. Why Radar?

I first came across radar when I joined the Navy, and can't say I was immediately impressed. The picture was quite big, and – so long as you kept your face squashed firmly into the rubber cowl over the screen – reasonably bright. But it didn't actually tell me very much.

Then, quite soon, things became much clearer: I found I could, after all, relate the glowing blobs on the screen to what I saw on the chart, and tell the difference between waves, buoys and ships. It wasn't very long after that that I found myself taking it for granted; even, on one memorable occasion, plotting a string of fixes by using the ship's big anti-aircraft radar to measure the ranges of mountain tops over 150 miles away.

That, though, was on a ship. Small craft radars were available, but they cost about the same as a small car, so it was some time before I got my hands on one. Almost at once, it proved its worth.

We were in southern Ireland. Maybe we were getting blasé about the weather, or maybe we simply didn't want to cut the cruise short. Whatever the reason, I left myself just two days to get home. With a cruising speed of six knots and 250 miles to go, we had no time to wait for ideal sailing conditions.

On the day we left, the early morning forecast gave us next to no wind, poor visibility, and fog patches. The land disappeared into the murk in a matter of minutes, but I was still plotting confident fixes by radar three hours later.

By nightfall, fifteen hours later, the visibility had closed in even more; for those on deck the world had shrunk to a circular patch of water not very much bigger than a football pitch. But in the relative comfort of the doghouse I'd found Bishop Rock, and soon afterwards I could see the low-lying Scilly Isles and even the Seven Stones lightship.

Still using the radar, we dodged the shipping heading north towards the Bristol Channel, and skirted the coast of south-west Cornwall.

Lizard Point showed up through the murk as a slightly darker patch of fog – but the fact that we could see it from a mile off showed that the visibility had lifted slightly. It could only have been a very temporary effect, though, because the next thing we saw was the stone beacon tower in the entrance to Falmouth harbour.

We could, perhaps, have made the same passage without radar. It is, after all, quite possible to maintain a reasonably accurate Estimated Position plot for well over two hundred miles. But it would take a braver man than I to slip through the six-mile wide gap between the Scillies and the Seven Stones with nothing better than a day-old EP to go on.

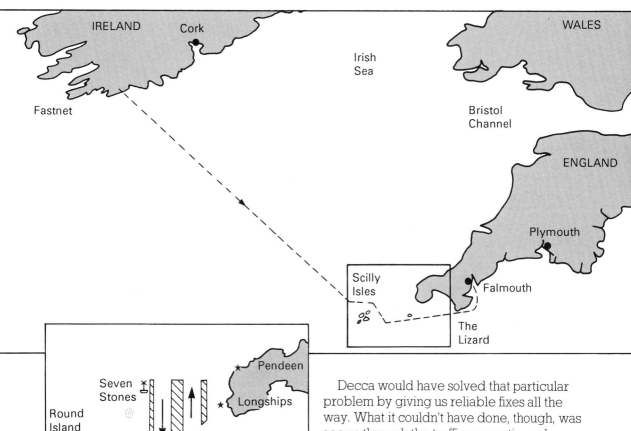

The route back from Ireland to Falmouth in the fog, showing the traffic separation scheme and navigational hazards in the Scillies area. The navigation, collision avoidance and pilotage were all achieved using radar; without it, the passage would have been virtually impossible.

Decca would have solved that particular problem by giving us reliable fixes all the way. What it couldn't have done, though, was see us through the traffic separation scheme. Nor could we have found our way into Falmouth with it.

So the radar made a potentially dangerous passage safe. It stopped a good cruise turning into a disaster, and in that one trip it probably saved enough in ferry fares to make up a fair proportion of its initial cost.

OK, so the versatility of radar makes it probably the most cost-effective of modern navigation aids. It's got something else in its favour, too. It's fun to use.

2. First Principles

Although a radar set is probably the most complex piece of equipment on board a small boat, its basic principle is startlingly simple: radio waves, just like light or sound waves, can be reflected by solid objects.

So if you transmit a short burst of radio waves, and get an echo back, you know, first of all, that there's something out there for them to bounce back from.

And if you know the direction from which the echo came, you know in which direction the object lies.

And thirdly – because radio waves travel at a constant speed of 300 million metres per second – if you measure the time between transmitting the burst and receiving the echo, you can work out how far the signal has travelled, and from that the range of the object. Suppose, for example, that there is one millisecond (0.001 second) between transmission and reception. At 300 million metres per second, or 162,000 nautical miles per second, that's enough for the signal to have travelled 162 miles: 81 miles out and 81 miles back.

◀ Radar transmits a 'pulse' – a block of radio energy travelling at 162,000 miles per second (1). When the pulse hits a solid object some of the energy is reflected back, still travelling at 162,000 miles per second (2). Meanwhile the rest of pulse continues to the horizon (3). The time that elapses between the transmission of the pulse and the return of the echo is directly related to the range of the object.

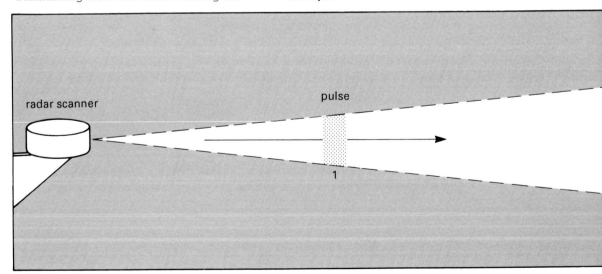

radar scanner

pulse

1

That, in essence, is how the first fully operational radars worked. They were set up in England in the mid 1930s to give early warning of air attack. Powerful transmitters operating in the 10-metre band (about 30MHz) broadcast over an arc of about 150°, floodlighting the area with invisible radio waves. Any aircraft entering the 'illuminated' area couldn't help but reflect some of the waves back to a pair of receiving aerials mounted at right angles to each other near the transmitter. The difference in signal strength in the two receiving aerials gave an indication of the direction from which the echo was coming, while an oscilloscope connected to the transmitter and receiver measured the time interval accurately enough to measure range.

It's obviously an enormous waste of power to flood a wide area with radio waves when you can only analyse an echo coming from one small target, so it would make sense to focus the power down a relatively narrow beam. Not only is this a more efficient use of power, but it also makes it much easier to measure bearing. You will only get an echo when the beam from the aerial is pointing at the target, so the transmitting aerial itself gives a direct indication of the bearing.

The snag is that to get a beam, say 3° wide requires an antenna 25 times as wide as the wavelength, so the 10m waves of the early 'Chain and Home' radars would have required antennas 250m across. One solution is to reduce the wavelength. Increasing the frequency from 30MHz to 10,000MHz reduces the wavelength to 3cm, making a 3° beamwidth possible with an antenna just 75cm wide.

Back in the 1930s that was simply not possible: there was no cost-effective or efficient way of producing these

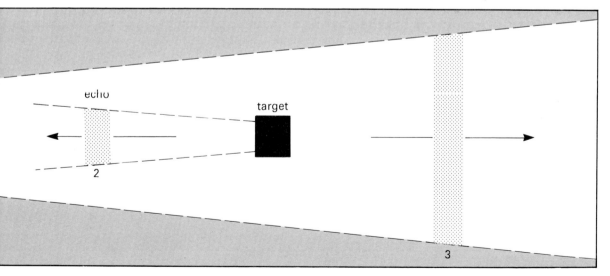

'microwaves'. But in 1940 a British team perfected the Magnetron: a compact, efficient, and relatively economical microwave generator.

Using microwaves brought other advantages besides efficiency. For one thing, higher frequencies make it possible to use much shorter bursts of radio waves. As we shall see, the shorter the pulse length, the better the radar is able to distinguish between two contacts close together.

Perhaps even more significantly, using the transmitting antenna to measure bearing led indirectly to a major change in the way radars presented their information.

The Chain and Home stations used two completely separate displays: one showing the target's bearing and the other – the oscilloscope – showing its range. In later displays, the spike which represented the returning echo on the oscilloscope was reduced to a blob. Meanwhile the trace, instead of remaining horizontal, was made to sweep round the screen in phase with the rotation of the antenna. This meant that range and bearing were shown together, on a single picture. It gave the operator the impression of looking down on a map or plan of his area, and gave us the name plan position indicator or PPI.

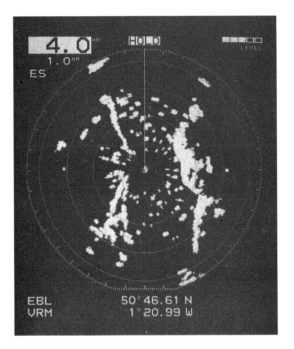

▶ **The radar image displayed on a PPI (top) shows the surrounding objects and coastlines in plan form, as seen from directly above the vessel equipped with the radar. This can be directly related to a chart of the area (bottom). Note the correlation between the coastline on the right of the chart and its representation on the radar screen.**

So by the middle of the Second World War, radar had developed to a stage at which it was directly comparable to what we have today. It had a transmitter, using a magnetron to generate pulses of microwaves . . . a rotating antenna, or scanner, which focused the outgoing transmission into a narrow beam sweeping the horizon . . . a receiver, which amplified the returning echo . . . and a plan position indicator, which displayed the range and bearing information received in an immediately intelligible form.

You don't actually need to know what goes on inside a radar set in order to use it. On the other hand, it's worth knowing what some of the terminology means to stop enthusiastic salesmen blinding you with science! For that reason, if no other, let's look briefly at each of these four main parts to see how each one contributes to its overall performance.

THE TRANSMITTER

The transmitter is almost invariably located inside the scanner unit, and its job is to create the microwave pulses. Within that very crude job description, though, there is scope for considerable variation. How powerful should each pulse be? How long should it last? And how long should the interval be between pulses?

Power

If the transmitter doesn't squirt out enough power then you're not going to get a discernible echo from distant targets or from poor reflectors, so transmitter power has an obvious influence on the effective range of the set. But it's a mistake to make too much of a virtue out of raw power, because there are other factors which influence operating range. Of these, the most significant is the fact that microwaves travel in almost straight lines, so a radar's view of the world is limited, just like your own, by a horizon.

Radar manufacturers seem more or less agreed on the optimum power output for a small boat radar, because they nearly all specify a nominal peak power output of between 1½kW and 5kW, with the vast majority in the middle of that range at about 3kW.

◀ **The radar transmitter is usually located within the scanner unit. In most small-boat installations this takes the form of a fully enclosed 'radome' which protects both the transmitter and the antenna.**

Pulse length

Pulse length is significant, because it too can have an effect on the useful operating range. A long pulse uses more energy than a short one of the same power, so any returning echo will include more energy and produce a more conspicuous contact on the screen. But there is something to be said for a short pulse length, too.

Imagine a radar pulse lasting 0.8 microseconds (less than one millionth of a second). By the time the end of the pulse emerges from the antenna, its leading edge will already have travelled 240 metres. Now suppose it travels outwards for a few miles until it comes across a pair of ships 100 metres apart. Part of the pulse will be reflected back from the nearest ship, but part will carry on to be reflected back from the other. Before the tail end of the pulse has reached the first ship, its leading edge will already be on its way back from the second one, so the echoes from the two ships will merge. So instead of two separate contacts, the ships will appear on the radar screen as a single large one.

Shortening the pulse length resolves the problem. By reducing the pulse length below the time it would take a radio wave to travel from one ship to the other and back, the ships appear as separate contacts.

So a short pulse length is required for good range discrimination, and a long one to detect weak targets at long range. To overcome these contradictory requirements, most transmitters are designed to operate at two or three different pulse lengths – typically about 0.5 microseconds for use at ranges over five nautical miles, and about 0.1 microseconds for shorter ranges.

➤ **Short pulse lengths improve range discrimination.**

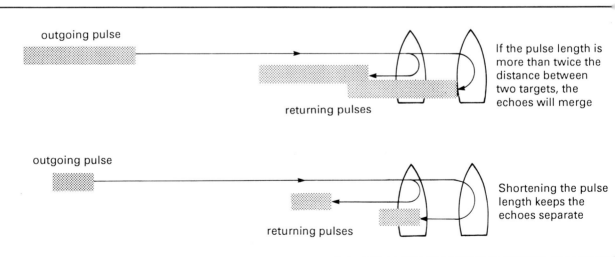

outgoing pulse

returning pulses

If the pulse length is more than twice the distance between two targets, the echoes will merge

outgoing pulse

returning pulses

Shortening the pulse length keeps the echoes separate

Pulse interval

The choice of pulse repetition frequency (PRF), the interval between pulses, is far simpler. The radar can't distinguish between the echo from one pulse and the echo from another, so the transmitter has to arrange things so that there is only one echo in the air at once.

If the set is operating to a range of 24 nautical miles, for example, an echo might be received at any time up to 296 microseconds after transmission. If it is operating on the 3-mile setting, any echoes that were going to come would have arrived within 20 microseconds. Ideally, perhaps, this would call for a different pulse repetition frequency for each range scale, but as most radars offer a choice of at least half a dozen different range scales this would be taking things a bit far. In practice, it is more common to have just two or three PRF's, one corresponding to each pulse length. For example, there might be 600 pulses per second at the long pulse length, and 2000 pulses per second at the short.

THE SCANNER

The antenna's job is to organize the outgoing microwaves into a tight beam. The simplest and cheapest way of achieving this is the so-called 'pill-box' antenna, made of thin sheet metal folded into the shape of a V, with a strip bent into a parabola filling the root of the vee. The parabolic strip serves to focus the microwaves in just the same way as the reflector of a torch focuses the light from a bulb.

◄ **The revolving antenna focuses the microwaves into a tight beam of radio energy.**

Most small marine radars now use a more sophisticated system in which the microwaves are fed into a hollow tube called a waveguide. This is sealed at both ends, but has a series of accurately-milled slots in one side. Each slot acts like a separate aerial, but their combined effect is to produce the required narrow beam from the aerial as a whole.

The most recent innovation, originally confined to military applications but now finding favour across a wide range of small-craft sets, is the patch aerial. This high-tech system uses a printed circuit consisting of an array of copper pads; these act as point-sources of microwaves roughly corresponding to the slots of a slotted waveguide.

Beamwidth

From the user's point of view, though, it is not how the antenna works that matters, but how well it works. The most crucial factor is the shape of the beam it produces. This is usually described in terms of horizontal and vertical beamwidth. Typical figures are about 4° and 25° respectively – implying, as the name suggests, that the beam of radiation is 4° from side to side and about 25° from top to bottom.

The term beamwidth is slightly misleading, because you might take it to mean that there is no power transmitted outside those limits. In reality, the typical situation is as shown in the diagram: the quoted beamwidth refers to the arc over which the power output is more than 50% of its maximum value. There is bound to be some power transmitted outside this arc, however,

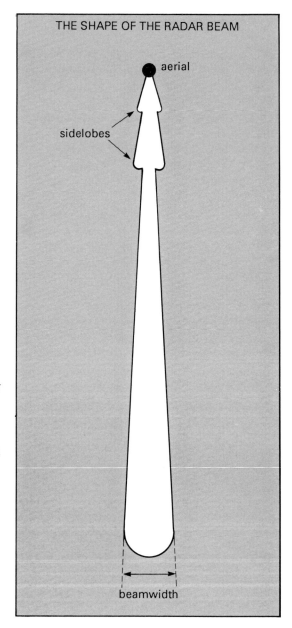

THE SHAPE OF THE RADAR BEAM

aerial

sidelobes

beamwidth

in particular in 'sidelobes' on each side of the main beam.

The effect of horizontal beamwidth can be compared to the effect of pulse length, in that it governs the radar's ability to discriminate between contacts which are close together. Suppose, for example, we have a small metal buoy, and a radar whose horizontal beamwidth is 4°. Because of the width of the beam, the first pulses to hit the buoy will do so when the antenna is still pointing a couple of degrees to the left of it, and they will go on hitting it until the antenna is pointing a couple of degrees to the right. In other words, the radar will 'see' the buoy through an arc of 4°.

If there were two buoys, so close together that the right-hand edge of the beam had reached the second one before its left-hand edge had cleared the first, the radar would not be able to distinguish the gap between them, but would show the two buoys as a single large contact.

So the smaller the horizontal beamwidth, the better. Vertical beamwidth is another matter altogether; a tight beam is definitely not required, because as the boat rolls the radar would find itself looking up into the sky and then down at the sea. Instead, the vertical beamwidth is deliberately kept up to about 25°, so that some part of the beam is always 'looking' horizontally.

◀ The shape of the radar beam.

◀ A narrow beamwidth improves bearing discrimination.

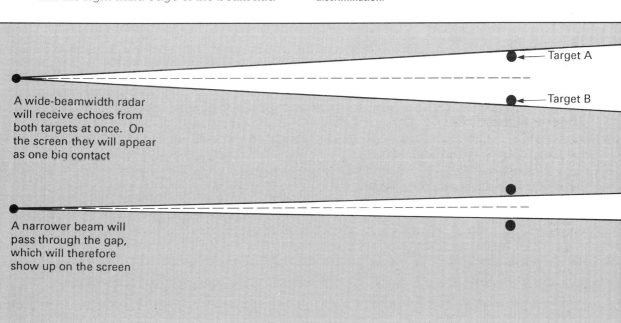

Target A

Target B

A wide-beamwidth radar will receive echoes from both targets at once. On the screen they will appear as one big contact

A narrower beam will pass through the gap, which will therefore show up on the screen

THE RECEIVER

Returning echoes are collected by the antenna: usually the same antenna as that used to transmit, but sometimes by a duplicate aerial bolted to it.

The echoes will, of course, be at more or less the same frequency as the outgoing pulses, and they must be converted to a more manageable intermediate frequency of about 60MHz.

Then they have to be amplified. This is a mammoth task because the echoes are inevitably very weak; the problem is also aggravated by the fact that the echo from a target twenty miles away will be over 99% weaker than the echo from a similar target a mile off. To cope with this kind of discrepancy, the degree of amplification applied to the signal varies, early returns being amplified by a factor of a few hundred and late ones by a million or more.

THE DISPLAY

Finally, the information is ready to be presented to the operator.

Until the early 1980s marine radars presented their information on an electro-mechanical (analogue) PPI. The picture was built up by an electron beam activating the fluorescent coating on the inside of a cathode ray tube, the beam sweeping out from the centre time and again, each sweep corresponding to the progress of a microwave pulse going out and back, and each successive sweep moving a little further round the screen in step with the rotating antenna. The picture, once created, lasted several seconds before fading, so it persisted until the scanner had completed a full circle and the picture was updated.

The result was a very clear, but very dim picture, which could only be viewed in darkness: the operator had to peer at the screen through a deep, light-tight cowl.

Then came raster-scan – a new generation of 'daylight-viewing' radars which use microprocessors to analyse the range and bearing information from the receiver and present it in a form more akin to computer graphics. Apart from the obvious advantage of being visible in broad daylight, raster-scan has made radar displays much more reliable, and has opened the way to all manner of increasingly sophisticated extra features. It is not, however, a wholly unmixed blessing.

In particular, some operators find their picture less clear than that of an analogue display. That's because the raster-scan picture is made up of thousands of tiny squares, called pixels (picture cells). So long as each pixel is small enough, that doesn't particularly matter – after all, the pictures in glossy magazines are made up of thousands of individual ink spots, but you'd never know it. Some cheap radars, however, use screens with relatively few pixels. These give a coarse, grainy picture in which every square is distinct, rather as the large spots used to print newspaper photographs make them look coarse and somewhat smudgy.

Another problem associated with some cheap raster-scan displays is that each pixel

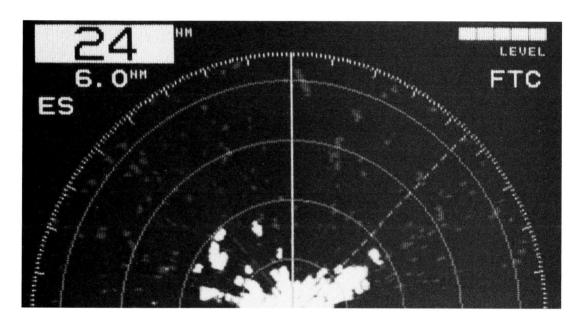

can only be either off or on. This means that there is no visible difference between a weak echo and a strong one: a contact either appears, or it doesn't, with no shades in between. Because of this, a very weak echo which would appear as a weak contact on an older set might not show up at all on a raster-scan display.

More expensive sets have quantized displays, in which each pixel can operate at any of several levels of brightness, reflecting the strength of the echo and allowing the operator to distinguish between a weak contact and a stronger one.

Some manufacturers have taken this a stage further with the introduction of colour radars, in which different colours, rather than different intensities, are used to indicate different strengths of echo. This undeniably

◆ **A raster-scan image consists of thousands of tiny pixels. In many sets these are either fully illuminated or blacked out, but some units employ a system of variable brightness to make the image more informative.**

makes the radar picture more attractive to look at, but personally I have yet to be convinced that it makes it any more informative.

Finally, there are a few liquid crystal displays (LCD). They create the picture using raster-scan technology, but instead of using a television-style screen, they use an array of liquid crystal diodes like those in a digital watch or portable computer. Their picture quality cannot hope to compare with even the simplest TV-type display, but they have the virtues of low price, compact size, and low power consumption.

3. Switching On

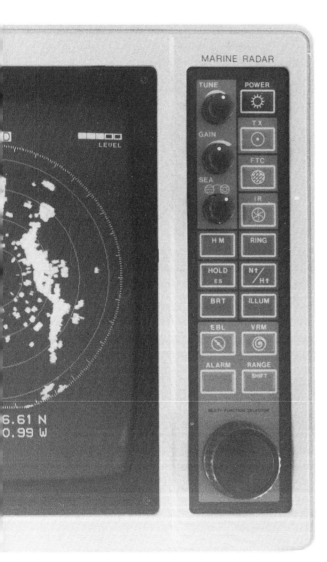

None of the frantic electronic activity going on inside a radar set is likely to be of immediate concern to the operator. He or she will be much more concerned with the question of how to manipulate the dozen or so controls on the front panel of the display unit.

Unfortunately, there's no such thing as a standard panel. Even the terminology varies from one manufacturer to another, and some even forsake written labels altogether in favour of symbols. The most important controls, however, are common to all radars, so although it may take a little while to get used to an unfamiliar radar, there is no reason why this should be any more difficult than getting used to an unfamiliar car.

Six controls in particular are more important than any of the others: between them, the on/off, standby/transmit, brilliance, gain, tuning, and range knobs determine whether you get a picture at all.

On, off and standby
The significance of the power on/off switch is obvious: if it is not switched on, the set is completely dead.

◀ **Control panels vary according to the manufacturer and the features of the set. This one uses a system of switches and a master adjustment knob.**

▶ **Switching on does not produce an instant image. The microwave generator needs to warm up, and then the set goes into 'standby' mode.**

But turning the power on is only part of the story, because the radar can't transmit until the magnetron has had a couple of minutes to warm up. Because of this, a separate switch may be included for the transmitter: alternatively, the on/off switch may have three positions, usually labelled 'off', 'standby', and 'transmit'. Apart from its main function of giving the transmitter a chance to warm up when the set is first switched on, the 'standby' mode is a useful way of saving power. In this state the radar is ready for instant use, but its power consumption is lower. The difference is not enough to concern motor boaters, but could make all the difference on a sailing yacht.

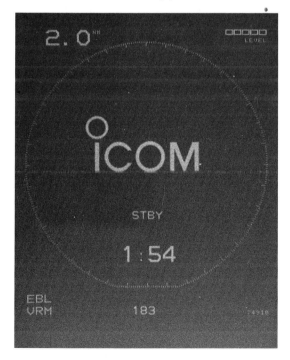

INTERNATIONAL SYMBOLS FOR RADAR CONTROLS			
anti-clutter rain minimum	anti-clutter rain maximum	anti-clutter sea minimum	anti-clutter sea maximum
scale illumination	display brilliance	range rings brilliance	Variable range marker
bearing marker	transmitted power monitor	transmit receive monitor	off
radar on	radar stand-by	aerial rotating	north-up presentation
ship's head up presentation	heading marker alignment	range selector	short pulse
long pulse	tuning	gain	

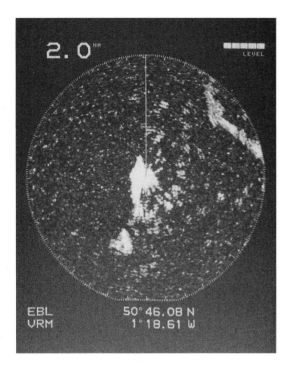

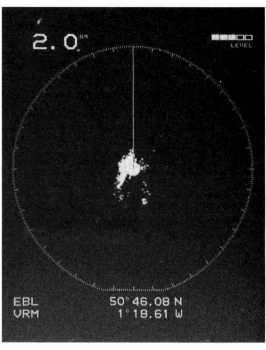

With the gain too high, the screen is covered with background speckle, making it very hard to read.

If the gain is too low the system fails to pick up the faint signals from distant contacts.

Brilliance and gain

The brilliance control regulates the brightness of the picture, making it bright enough to be seen in high levels of background light, or dim enough to preserve the operator's night-vision in darkness.

The gain control has a superficial similarity to brilliance, in that tweaking it makes the picture brighter or darker. Its function, though, is completely different, and it's important not to confuse the two. Unlike brilliance, which affects the display, the gain control regulates the receiver. Turning it 'up'

makes the receiver more sensitive, making weak echoes look stronger, but confusing the picture with background speckle – the visual equivalent of the background noise you hear on a radio when the volume is turned up too high. Turning it down – reducing receiver sensitivity – will eliminate the speckle, but if it's overdone the weak or distant contacts will disappear.

Tuning

The tuning control can also be compared to one of the controls of an ordinary radio or

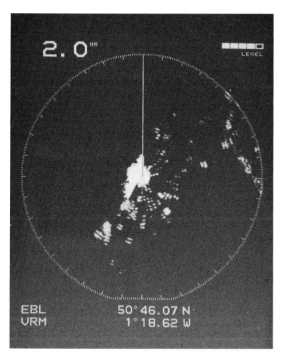

🔺 **With properly adjusted gain the distant contacts are plainly visible on a clear screen.**

television, in that it matches the receiver to the frequency of the transmitter. Of course, the radar receiver is 'listening' to signals which its own transmitter has sent, so it can be kept permanently tuned-in to the right frequency. But it is impossible to maintain absolutely perfect tuning over a long period, and the returning echoes are so weak that the tuning has to be spot-on if the receiver is to stand a chance. The knob on the front panel, therefore, is for fine tuning: coarse tuning will have been carried out either by the installation engineer or at the factory.

Range

The last of the main operator controls regulates the range at which the set operates. Most small-craft sets offer a choice of at least six range scales to choose from, typically from ⅛ mile to 16 or 24 miles. Electronically, changing the range affects several of the set's characteristics: its pulse length, pulse repetition frequency, and video presentation. So far as the user is concerned, it simply changes the size of the area covered by the picture, and hence the scale.

Just as you would use a different chart for pilotage than you would for passage-planning, so the choice of range scale depends on what you're using the radar for, and the area you're in. Groping up a narrow river in fog, for instance, you might choose a range of a half a mile to show plenty of detail close at hand. At sea, however, such a short range would be useless, because there shouldn't be anything that close that you don't already know about. So for coastal navigation you might well go out to a range of 16 or 24 miles to bring distant landmarks into the picture. For collision avoidance you'd choose an intermediate setting of perhaps 8 or 12 miles.

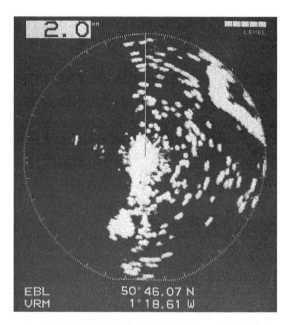

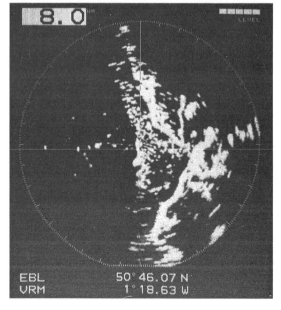

▶ **A short-range image with the radar switched to a radius of two nautical miles (top) and a medium-range image covering a radius of eight nautical miles (bottom). The chart on the far right shows the area covered in each case. In these crowded waters the mass of contacts representing other vessels become quite unreadable on the longer range scale.**

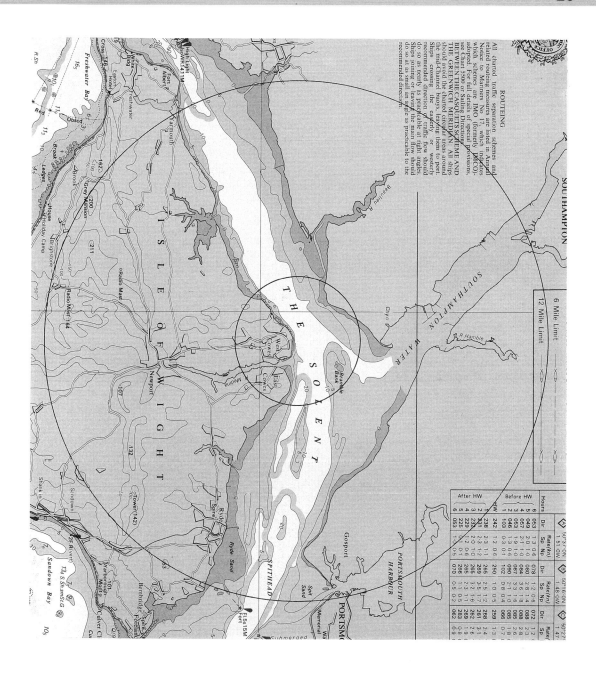

THE START-UP ROUTINE

The first move in switching on a raster-scan radar is obvious: turn the power on. If, however, you're using an older set, with an analogue display, you should first turn the brilliance and gain controls right down. That's because the way in which the picture is produced involves squirting a beam of electrons at a delicate fluorescent coating inside the screen. When the set is first switched on the beam is concentrated at the very centre, and would, over a period of time, cause irreparable damage to the coating. Turning the brilliance and gain down reduces the intensity of the beam, and thereby minimizes the damage to the screen.

Then you have to wait, while the transmitter warms up. Some sets give a countdown: others keep you in suspense, just displaying the word 'wait' – or something to that effect – until the time is up. Others show an indicator light on the panel when the set is ready. Once the warm-up process is complete, the set will be in its standby mode, and the transmitter can be turned on whenever you're ready.

As soon as the set is switched from standby to transmit, you should get a picture, of sorts, but it can almost certainly be improved.

Start by adjusting the brilliance to suit the conditions. Bear in mind that on an analogue display the brilliance should have been turned right down before switching on, and that most raster-scan displays automatically set themselves to a fairly dim picture when they're switched on. The pattern of

concentric range rings is the thing to look at: they should be distinct, but not dazzlingly bright or blurred.

Next, adjust the gain, turning it up until a light background speckle appears, then turning it down again until the speckle just – but only just – disappears.

By this stage you should have something in the way of a picture. If the set is switched to its maximum range, though, and you're in a confined space such as a marina, you may well find that the whole picture is concentrated into a pinhead-sized blob in the middle of the screen. So your next move has to be to choose a suitable range.

Ideally, for the final stage in the process, this will be one of the middle ranges of those on offer. Somewhere round about four miles is usually recommended, but the most important thing is that there has to be at least one reasonably distant target in range, even if it's not, at this stage, visible on the screen.

That's because the last job is tuning, and in order to tune effectively you have to be receiving something. If possible, pick a distant target because the effect of tuning is more obvious on a weak contact than a strong one, but it's not always possible to be so choosy. Most sets have a tuning indicator, which can simplify the process, but you still need a target within range.

Tuning properly only takes a couple of minutes, so don't rush it. Move the tuning control in small steps, and wait at least three seconds after each move to see what effect it has had on your chosen contact (or the tuning indicator) before trying it again. Eventually you should find one particular

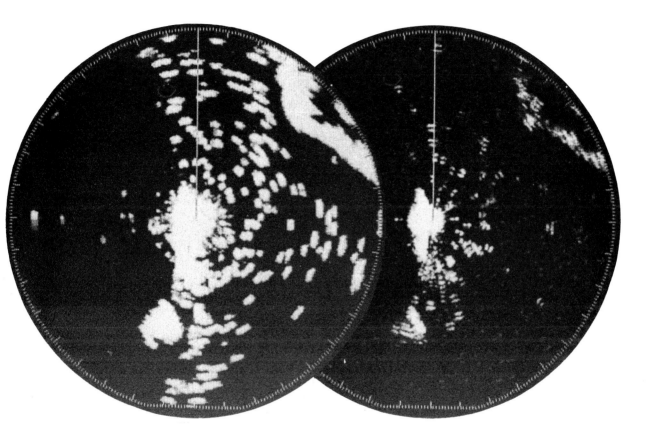

setting which gives a clearer, brighter picture than any other, with more contacts in view. That's it, for the time being. It's worth checking it again though, after the set's been working for half an hour, and again after every hour or two of operation.

It's easy to remember the start-up sequence: once you're switched on, adjust the **B**rilliance, **G**ain, **R**ange and **T**uning in alphabetical order.

◆ **When the set is properly tuned (left) the contacts show up bright and clear all the way to the edge of the screen. If the set drifts out of tune (right) you get a fainter image, and many of the more distant contacts may fade out completely.**

4. Interpreting the Picture

Less than five years ago I could have started this chapter with something like 'The vital thing to remember when you look into a radar screen is that your own boat is always at the centre, and heading upwards.'

That's not true any longer: a fair proportion of recent models allow you to move your own boat away from the centre, and some can turn the whole picture round to put north at the top. We'll look at these variations in chapters 9 and 10, but for the time being we'll stick to the standard 'head-up' centred display, because apart from being the most common, it's also the most useful.

Having your own vessel at the centre means that the picture on the screen is rather like looking down on the area around you from a great height, seeing the land, buoys, and other vessels spread out as though they were on a map. And having the head up makes it easy to relate the real world to the radar image, because it means that anything on the right-hand side of the screen is to starboard, and anything on the left-hand side is to port.

To further help you orientate yourself, the 'straight ahead' direction is marked by a straight line pointing upwards from the centre of the screen, called the heading marker, and a series of equally-spaced concentric range rings are superimposed on the picture to help estimate range. The number and spacing of the rings varies from

set to set and depends on the range scale in use: typically there might be four, six or eight rings on the longer ranges, and two or three on very short ranges.

Look carefully at the radar picture and at a chart, and you'll soon be able to match up certain features – headlands, or distinctively-shaped stretches of coastline – which show up on both. Buoys, so long as they are within a range of about four or five miles, should

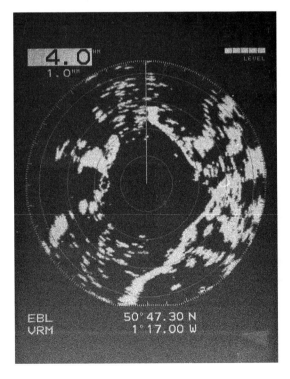

appear as blobs or specks, as will ships or other boats.

Almost inevitably, though, there will be some discrepancies between the chart and what you see on the screen. Some things on the chart won't appear on the screen; others may be on the screen but not on the chart; while others will show up on both, but with their size or shape distorted.

So interpreting the picture takes some practice.

☛ The image on a 'head-up' display (left) is related to the heading of the boat – represented by the heading marker pointing straight up from the centre of the screen. The chart (below) shows the area covered.

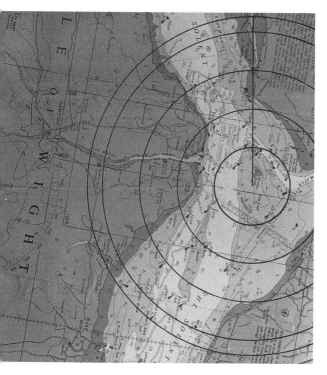

WHAT IT CAN'T SEE

Remember that in order to produce a contact on the screen, a target first has to be 'illuminated' by the microwave pulses from the antenna, and then has to reflect them back to the receiver.

Various things can conspire to prevent pulses ever reaching their target: first, and most obviously, solid obstructions on the boat itself. Imagine a radar set mounted on the forward face of a steel funnel. Any pulses directed towards the funnel will hit it, and will bounce straight back. Microwaves travel in straight lines, so the funnel will create an area astern over which the radar is completely blind.

Checking for blind arcs is a quick and easy process. All you need is a patch of slightly choppy water!

Start by switching down to a range of about half a mile or a mile, making sure that the 'sea clutter' control is turned right down or switched off. Most of the screen will be filled with a mass of small contacts, caused by the echoes returned from waves. Any dark streaks radiating outwards from the centre represent areas where there is no sea return – in other words, blind arcs. It's worth marking these on the edge of the screen as a reminder of where they are.

A similar effect occurs when a substantial object comes between the antenna and a target. The classic example is a headland: radar can't 'see' through such a solid object any better than you can, so anything beyond a prominent headland or large ship will be hidden in a shadow area.

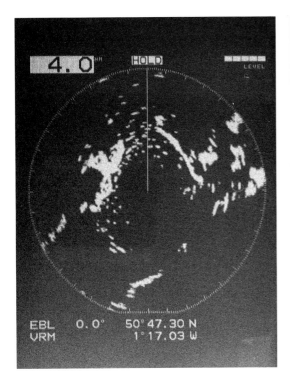
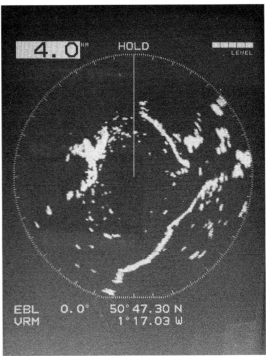

⬆ The radar image above is suffering from a blind arc caused by an obstruction on the boat. It is virtually blank on the bottom right-hand side, yet this is not obvious until the obstruction is removed (right) and the coast reappears to form a continuous line.

The most substantial object of all, of course, is the earth itself, which limits our own eyesight to the visual horizon. It has just the same effect on radar. Microwaves are refracted (bent) very slightly more by the atmosphere than are light waves, so the radar horizon is a shade further away than the corresponding visual horizon, but it still imposes a limit on the effective range of even the most powerful radar.

The horizon can produce some strange effects, particularly when you are making a landfall on an uneven coastline. Hills will show up before the rest of the land, so a long stretch of coast may look like a string of islands to a radar set well offshore. Then, as you get closer, the islands may merge to form what appears to be a solid shoreline, but bearing no resemblance to the coast shown on the chart.

Despite all this, a well adjusted radar is better at spotting buoys than the human eye, because it will see a buoy as soon as its radar reflector appears above the horizon; the naked eye needs a much bigger target.

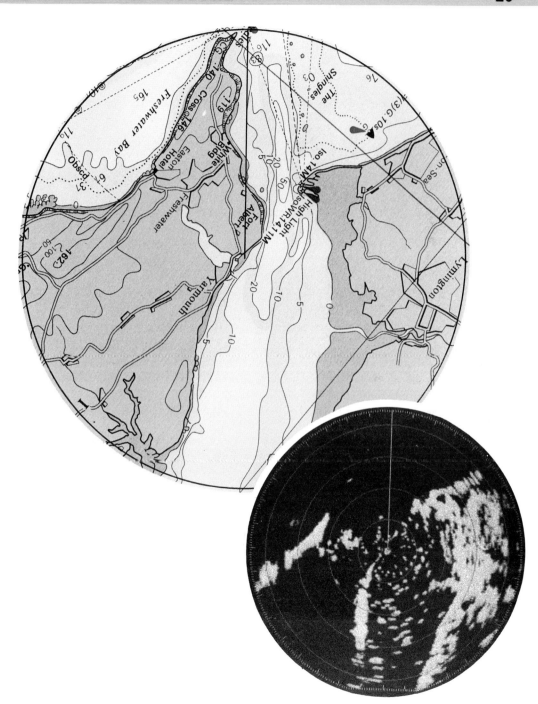

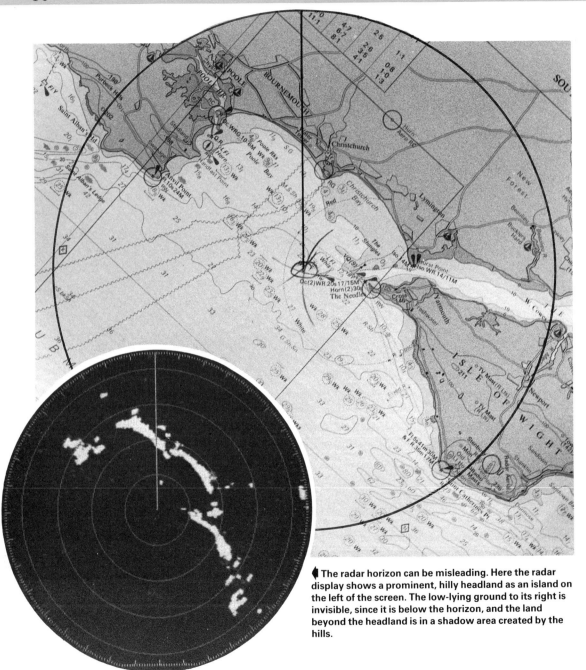

◖ The radar horizon can be misleading. Here the radar display shows a prominent, hilly headland as an island on the left of the screen. The low-lying ground to its right is invisible, since it is below the horizon, and the land beyond the headland is in a shadow area created by the hills.

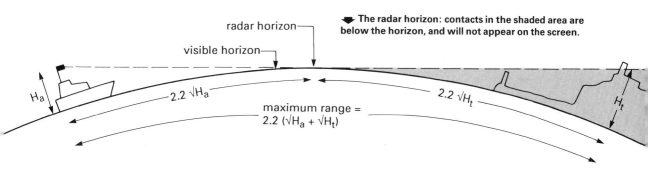

radar horizon

visible horizon

☛ The radar horizon: contacts in the shaded area are below the horizon, and will not appear on the screen.

H_a

$2.2 \sqrt{H_a}$

$2.2 \sqrt{H_t}$

maximum range = $2.2 (\sqrt{H_a} + \sqrt{H_t})$

H_t

It's possible to work out the maximum range at which you could expect to detect a target using a simple mathematical formula:

$$R = 2.2 (\sqrt{Ha} + \sqrt{Ht})$$

where R is the range in miles, Ha is the antenna height in metres, and Ht is the target height in metres. For most practical purposes, though, it's just as effective to use ordinary 'dipping distance' tables, but add about 5% to allow for the extra refraction.

WHAT IT CAN SEE

Once a microwave pulse has reached a target, it then has to be reflected back. Some things are better at doing this than others, depending on their size, shape, orientation, surface texture, and material.

The effect of size is pretty straightforward: all other things being equal, a big target will reflect more of the microwave energy than a small one, so it will produce a stronger echo, and show up on the screen as a brighter contact or at longer range.

The effect of shape is less readily explained, but it can be visualized by thinking of each radar pulse as a ball. It is considerably easier to hit a ball accurately with a tennis racquet than with a baseball bat, because the tennis racquet has a more or less flat surface. In contrast, the round surface of a baseball bat means that the ball has to be struck absolutely true if it is not to ricochet away unpredictably. Generally speaking, then, flat surfaces are better reflectors than curved or pointed ones.

☛ Rounded objects tend to scatter radar waves instead of returning them.

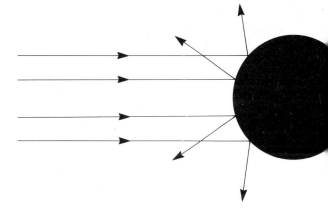

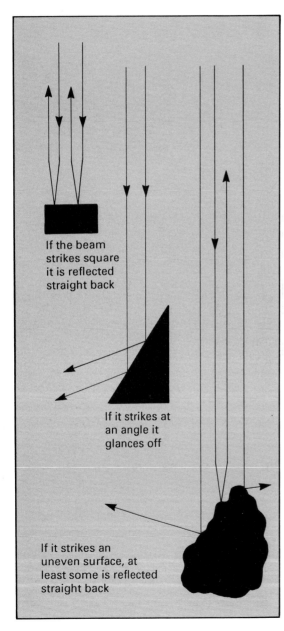

If the beam strikes square it is reflected straight back

If it strikes at an angle it glances off

If it strikes an uneven surface, at least some is reflected straight back

Orientation is really a variation on the same theme, because the reflected pulse has to be sent back to its source. In other words, it's no good having a superb reflective surface if it shoots the microwaves off in completely the wrong direction. So a ship broadside-on will produce a much brighter echo than one which is at even a slight angle.

From this it might seem that a good reflector at an angle to the line of the radar pulse would produce no contact at all, bouncing the pulse off in the wrong direction. The saving factor is scattering, caused to a large extent by surface texture. A perfect, mirror-smooth surface would indeed produce a weak contact unless it were perfectly aligned at right angles to the pulse. In real life very few reflectors are that good: almost everything includes surface irregularities which scatter the pulse, so that at least some of it goes in the right direction. Generally speaking, therefore, rough textures produce more reliable echoes than smooth ones – not quite as strong, perhaps, but less dependent on orientation.

Finally, there's the big one – material. It's well known that wooden boats are almost invisible to radar. That's because wood absorbs microwaves instead of reflecting them. GRP, too, is a very poor reflector, because microwaves go straight through it.

Rock and concrete are far better, and because they tend to occur in fairly large chunks they can usually be relied upon to produce a good strong contact.

Metal, though, is undeniably the best, but bear in mind that the small size and

⬆ **A quite substantial wooden or GRP boat can be virtually invisible to radar unless it has a well-designed, properly mounted radar reflector.**

cylindrical shape of a metal mast are likely to count against it, so it's bad policy to believe that if your boat has a metal mast you don't need a radar reflector.

RADAR REFLECTORS

⬆ **A brace of radar reflectors: the traditional octahedral type (left) and a modern encapsulated model (right).**

Many boat-building materials such as wood and GRP are almost invisible to radar. If such craft are to show up on the radar screens of other vessels they must be equipped with adequate radar reflectors. There are two basic patterns: the traditional octahedral reflector and the encapsulated type.

Octahedral reflectors are fine as long as they are mounted properly in the correct 'catch-rain' position. This can be difficult unless you use a rigid mounting as shown above.

An encapsulated reflector can be hauled up the mast on a halyard, but it is less easy to mount on a motor boat.

RADAR MIRAGES

Atmospheric conditions and reflections can play tricks on your eyes. The radar can also be fooled into 'seeing' things that aren't really there, or into seeing them in the wrong place.

Perhaps the commonest of these effects are caused by sidelobes – relatively weak beams of microwave energy on each side of the main beam. Modern antennas have made it possible to reduce sidelobes to such an extent that they can be ignored most of the time. They can be significant, however, if you are operating close to a particularly good reflector such as a large ship.

The microwaves from the sidelobe are reflected from the ship just like those from the main beam, and are received by the radar. It has no way of knowing that the echoes have come from a sidelobe, so it displays them as though they had come from the main beam. The result is a distinctive pattern of weak contacts on the screen, on each side of the genuine contact, and at exactly the same range.

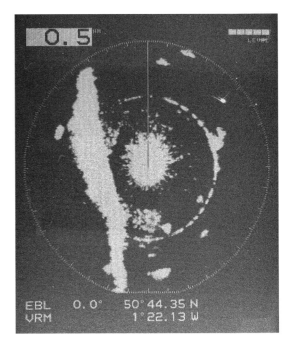

▲ The bright ring on this radar screen is produced by sidelobe echoes reflected from the ship pictured below. The ship itself is the strong contact just above centre on the right of the screen; the sidelobe echoes produce mirages at exactly the same range, forming a ring. The broad streak on the left is the coast.

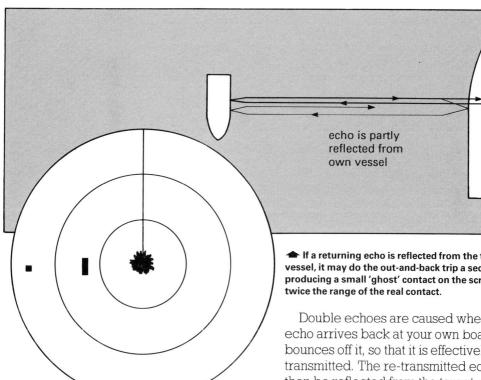

echo is partly
reflected from
own vessel

⬆ **If a returning echo is reflected from the transmitting vessel, it may do the out-and-back trip a second time – producing a small 'ghost' contact on the screen at exactly twice the range of the real contact.**

Double echoes are caused when a strong echo arrives back at your own boat and bounces off it, so that it is effectively re-transmitted. The re-transmitted echo can then be reflected from the target a second time, and this second echo may be received by the radar.

The chances of this happening are remote unless both your own boat and the target are particularly good reflectors: when it does, it results in a second contact on the same bearing as the first but at exactly twice its range.

Reflected echoes are also caused by a radar pulse bouncing from a good reflector. In this case, the reflected pulse goes on to hit another target, and is then reflected back to the antenna. This is probably the most difficult to identify,

Most other radar mirages are caused by pulses from the main beam being reflected so that they arrive at the antenna from the wrong direction or later than they should. They aren't exactly everyday effects, so most of the time they pose no problem. On the other hand, because they appear so rarely, they can be very confusing when they do happen. They're worth bearing in mind when you're operating a radar close to a large ship, a bridge, or a harbour wall.

because by its very nature the false contact can appear almost anywhere on the screen.

A reflected echo can produce a contact in a blind arc, or in the shadow zone beyond a headland or large ship. Perhaps the most distinctive feature of reflected echoes – or 'ghosts' – is that they usually appear suddenly at short or moderate range, then disappear just as abruptly.

Reflection can even be caused by a part of your own vessel, and produces exactly the same effect except that the ghost contact will almost invariably be in one of your radar's blind arcs.

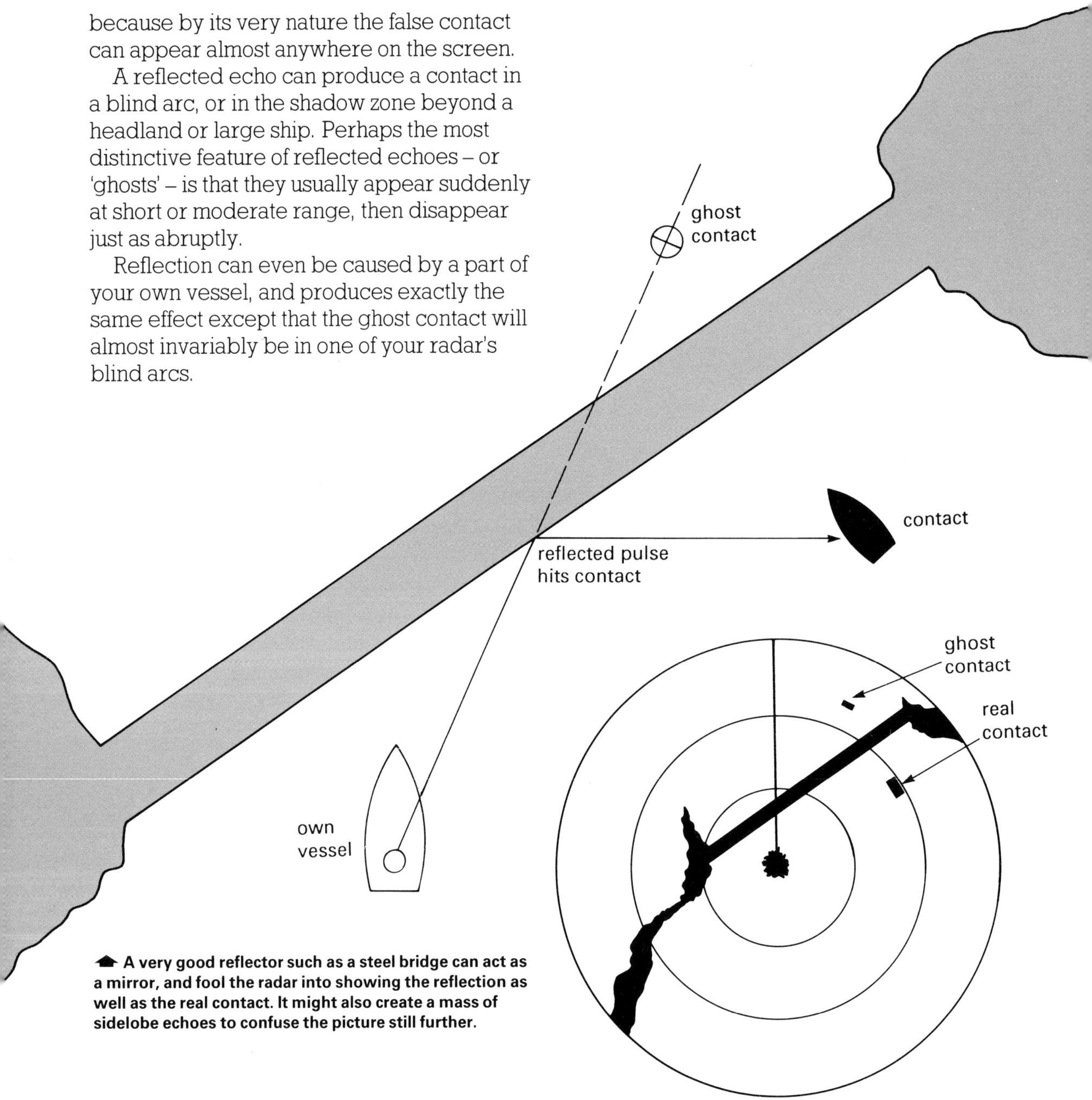

ghost
contact

contact

reflected pulse
hits contact

ghost
contact

real
contact

own
vessel

A very good reflector such as a steel bridge can act as a mirror, and fool the radar into showing the reflection as well as the real contact. It might also create a mass of sidelobe echoes to confuse the picture still further.

DISTORTION

A well set up radar won't bend things out of shape, but its lack of discrimination can make things appear differently to the way they really are.

The most common type of distortion, affecting every single contact on the screen, is directly related to the radar's beamwidth. In Chapter 2 I mentioned that a set with a beamwidth of 4° will show a buoy as a contact 4° across. In fact, it will enlarge all contacts, extending each side by an angle equal to half the beamwidth.

So an island 10° wide would appear to be 14° wide, while a headland would appear to extend 2° further to seaward than it really does. The most dramatic example of this occurs when you're looking at a narrow passage between two obstructions: from a distance the two will appear to merge together.

If it's really important to distinguish the gap, the effect can be reduced by turning down the gain, but remember to turn it up again afterwards.

Range discrimination, described in Chapter 2 as being related to pulse length, can have similar effects. Generally speaking it's less significant, but it's still worth keeping at the back of your mind, because a rock which looks like an easily identifiable feature on the chart could be no more than a blip on the coastline when you see it on the radar. It's difficult to overcome this particular problem, other than by knowing that it exists and making allowances for it. A noticeable improvement can be achieved by switching to a shorter range scale, or, on some sets, by switching to a shorter pulse length.

5. Improving the Picture

Apart from their essential controls, most radars have a selection of additional features intended to improve the picture.

Two of these in particular are so useful that they are now virtually standard: sea clutter (sometimes called STC, anti-clutter sea, or swept gain) and rain clutter (sometimes called FTC, anti-clutter rain, or the differentiator). As their usual names suggest, both are intended to clarify the picture by removing clutter – getting rid of spurious echoes received from waves and rain storms. Despite the similarity in their names and function, however, they work in completely different ways.

SEA CLUTTER

Sea clutter shows up as a bright sunburst pattern in the centre of the screen, caused by the echoes received from the sea surface. Waves aren't especially good reflectors of microwaves, so the range at which a wave is likely to show up is limited – and could be reduced by turning down the gain. The snag with that solution, of course, is that more distant contacts may be lost as well.

The sea clutter control is a refinement of this idea. It reduces the receiver gain for a few microseconds after each pulse has been transmitted, then gradually restores it to its normal level. It's very effective, and can usually eliminate sea clutter altogether, but using it calls for some care. Overdo it, and you could end up throwing the baby out with the bathwater, losing important close-range contacts as well as the unwanted ones.

▶ You need to use the sea clutter control with caution. With the control set low (left) there is a lot of clutter at the centre of the screen, but the nearby contacts – the ones that might collide with you – show up well. Turning the control up too high eliminates the clutter, but eliminates all the nearby contacts as well.

RAIN CLUTTER

Raindrops are even worse reflectors of radar pulses than waves, but there are an awful lot of them. In a rain, hail or snow storm the cumulative effect of millions of tiny echoes can be strong enough to register on the screen, producing a mass of weak contacts spread across a wide area. The effect on the screen is usually no more than a diffuse smear, often described as looking like a patch of cotton wool, but it can be dense enough to conceal more solid echoes.

Because rain clutter can appear anywhere on the screen, the gain reduction approach is useless. Instead, the rain clutter circuitry makes use of the fact that the echo from a rain storm is significantly different from the

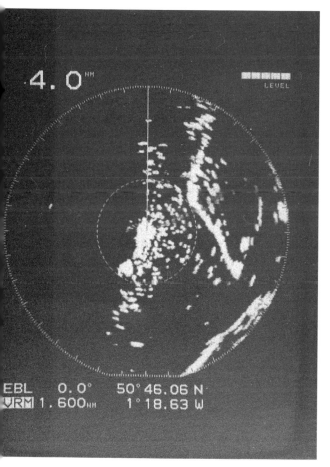

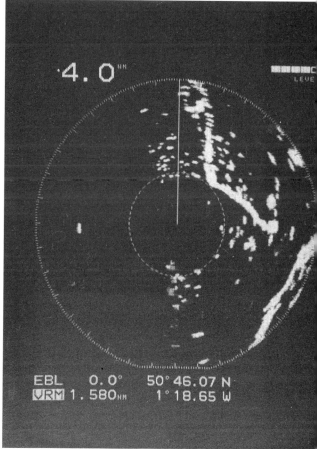

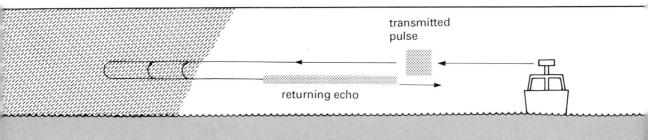

echo produced by a solid object: being made up of thousands or millions of tiny echoes instead of one strong one it's much longer but very much less intense.

The rain clutter control switches an auxiliary circuit into the receiver which clips each echo so that only its leading edge is passed on to the display. In the case of a crisp, strong echo such as you might get from a ship, this has relatively little effect. Drawn-out echoes, however, will be weakened, so the already weak echo from rain, hail or snow may well be lost altogether.

◆ Because a rainstorm is made up of a mass of very small reflectors – raindrops – a tiny proportion of the radar pulse is reflected from innumerable points throughout the storm. The result is a weak but highly extended echo.

◆ The rain clutter circuit clips each returning echo so that only its leading edge remains. In the case of a weak, extended rain echo, the signal that is left is too slight to show up on the screen. With a normal echo the signal that is left is still big enough to register – although it is fainter than usual.

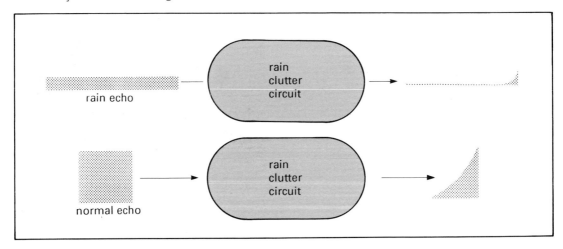

In practice, very few echoes are so crisp as to remain completely unaffected, so switching in the rain clutter circuit usually produces an across-the-board reduction in contact strength. This is particularly noticeable with images of the land, which become much fainter and take on a speckled appearance.

INTERFERENCE REJECTION

A third type of clutter is caused by other radars. Although no-one else is likely to be operating on exactly the same frequency as you, the transmissions from another radar nearby are so much stronger than the faint echoes your receiver is designed to detect that they will be picked up anyway.

On short-range scales the effect is seldom particularly serious. Only a small proportion of the intruding pulses will show up on the screen, where they will appear as an apparently random scattering of intermittent pinpricks of light.

On long-range scales, more of these pinpricks will appear, and they will form a distinctive pattern of dotted lines curving outwards from the centre of the screen. Even this isn't too serious so long as only one other radar is involved, particularly as the effect will disappear once the distance between you and the other vessel increases. In busy

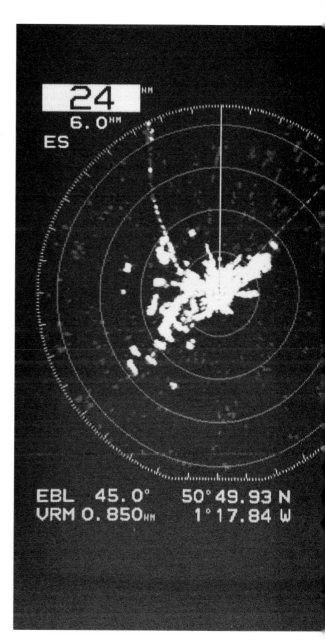

▶ **The bright arc curving up from the centre of this screen is a typical example of radar interference generated by the powerful transmissions of another radar set. Normally such effects are easily recognised and fade rapidly.**

areas, though, where there may be several vessels in close proximity, the clutter can be dense enough to cause confusion, and is less likely to go away of its own accord.

An interference rejection circuit can solve the problem, by rejecting any 'echo' which does not return from each of two successive pulses. This requires fairly sophisticated electronics so it's a relatively recent development so far as small boat radars are concerned, but it's rapidly becoming a much more common feature.

Unlike the sea and rain clutter controls, interference rejection produces little in the

way of unpleasant side-effects. Fortunately – and in spite of the advice given in some manufacturers' manuals – it doesn't eliminate the one type of outside interference which is a real benefit: racons.

Racons are automatic radio beacons transmitting on radar frequencies and fitted to many of the most significant navigation marks. When a racon is triggered by receiving a stream of radar pulses, it

➤ **Here the ship's radar has activated a racon directly ahead, generating the bright streak extending up the heading mark at the top of the screen. The nearest point of the streak is the position of the beacon.**

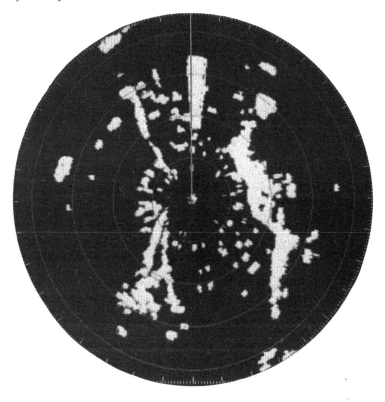

responds by transmitting a signal of its own, on the same frequency as the triggering radar. This appears on the screen as a bright streak, extending radially beyond the contact produced by the echo from the mark, and serves to identify the mark as well as drawing attention to it.

EXPANSION

Ships may well have one person on watch whose sole job is to keep an eye on the radar. In small boats the radar operator will almost invariably be keeping a visual watch as well, and possibly navigating and steering into the bargain. To make it easier to spot weak contacts, many small boat radars have the ability to expand them.

Variously named expansion, echo stretch or pulse stretch, its sole object is to make small contacts look bigger. This can be a great help, but it does tend to make the picture look very 'blobby' and indistinct, reducing range discrimination. It's a good idea to switch the expansion off when using radar as a navigation aid rather than for collision avoidance.

6. Navigating by Radar

Unlike most other electronic navigation aids, radar does not give a direct indication of latitude and longitude: getting a fix from radar is more akin to traditional navigation methods in that you have to identify landmarks, measure their position relative to your own, and use those measurements to construct two or more intersecting position lines on a chart.

This is less complicated and time-consuming than it sounds. Taking measurements by radar is often far easier and more accurate than using a hand-bearing compass, and plotting the fix may well be quicker than transferring latitude and longitude to the chart from a Loran or Decca set.

The accuracy of the fix depends on several factors, including the equipment itself and the ability of the operator. Significantly, though, accuracy invariably improves as the operating range decreases, so you get good fixes when you most need them – close to land.

FIXING BY RADAR BEARINGS

Anyone used to traditional navigation tends to think of visual bearings as the primary method of taking a fix. Radar can be used to take bearings too – without the limitations imposed by darkness or poor visibility.

The first requirement, of course, is to pick landmarks which are recognizable on the screen as well as on the chart. The flagpoles, monuments, castles and churches used for visual fixes rarely show up on radar, but headlands, islands, buoys and beacons do, and are often more easily identified by radar than by eye.

If you have a choice, the same considerations apply as when choosing

◄ The area covered by the radar shown opposite. The first step is to find a landmark that is recognizable on the screen – in this case Calshot Spit.

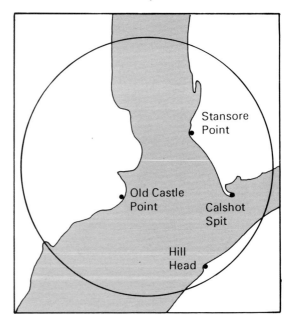

Stansore Point

Old Castle Point

Calshot Spit

Hill Head

visible landmarks: avoid using features whose bearings are within 20° or 30° of each other or nearly reciprocal, and use near features rather than distant ones. This is particularly important when using radar, because although no-one would think of taking visual bearings of something that's below the horizon, radar can seduce you into thinking that you're looking at a shoreline when what you're actually seeing is the top of a hill somewhere inland.

Using the EBL

Radar bearings are usually taken using an electronic bearing line, or EBL – a straight

TRUE? COMPASS?

line from the centre of the screen to its edge, which can be moved around the screen at will. To take the bearing of a buoy, for example, you move the EBL so that it cuts through the centre of the buoy on the screen, and read off the bearing from the EBL readout in one corner of the screen.

Older (analogue) sets don't have an EBL. Rough bearings could be taken using a scale of degrees engraved around the edge of the tube, but more up-market versions have a mechanical cursor: a transparent screen mounted in front of the tube and rotated by hand. An engraved line on the cursor screen serves exactly the same purpose as an EBL.

➤ The EBL has been activated and aligned with the edge of the spit. Note the bearing displayed at the bottom left of the screen: 112.5°. *SHIPS HEAD ? ? ?*

➤ As the boat yaws the whole picture swings round, and you have to chase the contact round the screen with the EBL. The bearing is now 71.5°.

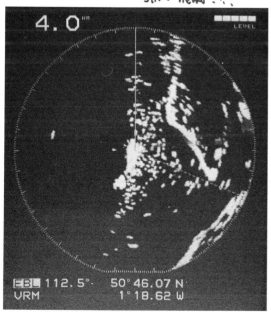

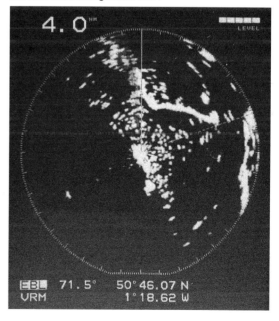

Some of the latest sets, while still equipped with an EBL, are rendering it almost obsolete by incorporating an electronic cursor. This isn't a line, but a cross, which can be moved around the screen by a keypad of up/down/left/right buttons or a 'tracker ball' until it coincides with the chosen contact. A numerical display in one corner of the screen gives the range

[handwritten annotation: Plan on USL up North up | SHPS HdH | EBL | MGNETU |]

To obtain a fix using radar bearings, you need to align the EBL with at least three recognizable landmarks, noting the heading of the boat as you take each bearing. The sequence below shows, left to right, a bearing on Calshot Spit (90.5°), Hill Head (145.5°) and Stansore Point (9.5°). The relative positions of these landmarks are shown in the sketch on page 44. On the far right is a photograph of the chart with the fix marked upon it. The rather large 'cocked hat' indicates the limited accuracy of radar bearings.

and bearing of the cursor from the centre of the screen.

Employing a radar bearing

Whichever method you use, taking radar bearings is usually far quicker than doing the same job with a hand-bearing compass. Actually taking a bearing, though, is only part of the job.

Remember that most small-craft radars have 'head-up' displays, on which your own heading is represented by a heading mark pointing straight upwards. So any bearing taken from a head-up display will be 'relative' – that is to say it will be related to your ship's heading, rather than to the compass.

Converting a bearing from relative to compass is a matter of straightforward

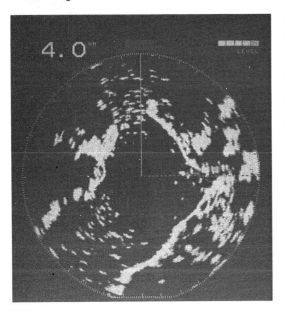

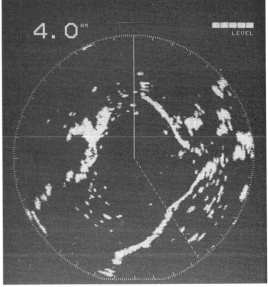

arithmetic: simply add the relative bearing to the heading. For example:

Relative bearing	047°
Heading	+ 190°
Compass bearing	237°

If the result is more than 360°, subtract 360° to get a practical answer. For example:

Relative bearing	286°
Heading	+ 190°
	476°
	− 360°
Compass bearing	116°

Bear in mind that if your steering compass is magnetic, the answer will still have to be corrected for deviation and variation before it can be applied to the chart.

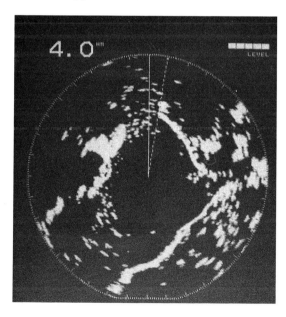

Sources of error

The conversion from relative to true may be simple, but it's probably the biggest source of potential error, because the accuracy of the whole process depends on you knowing the exact heading of the boat at the moment the bearing is taken. A 5° yaw at the wrong moment will produce a 5° error in the bearing.

There are two ways round this, of which the most obvious is to make fixing a two-man job, the helmsman noting the actual heading as soon as the navigator has lined up the EBL.

Those whose sets have two EBLs can use a more sophisticated technique, lining up the first EBL with a contact somewhere ahead or astern while the boat is on course. The navigator can then see immediately if the boat yaws, because the contact will wander off to one side of the EBL. The second EBL, meanwhile, is used to take bearings.

A second source of error is inherent in the system itself, because it is related to the beamwidth of the set. This, you will remember, will make a small target such as a buoy look as though it is several degrees across, and extend each side of a larger contact by an angle equal to half the beamwidth.

So if you are taking a bearing of a small contact, the EBL should cut through the middle of it, rather than brushing one edge. Large contacts, such as headlands, are more complicated: the standard advice is to move the EBL so that it passes just inside the contact by a distance equivalent to half the beamwidth. That's probably the best you can do, but bear in mind that adjusting the gain or tuning, or switching in the rain clutter or expansion circuitry will make a difference to the apparent size of the contact. Weak contacts will also be enlarged less than strong ones.

◀ **If you are taking a bearing with the EBL, ask the helmsman to call down the ship's heading at the moment you note the bearing on the display.**

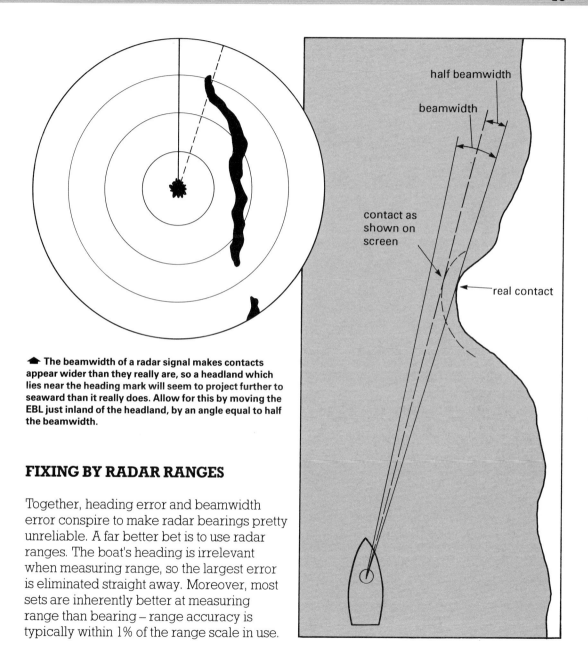

⬆ **The beamwidth of a radar signal makes contacts appear wider than they really are, so a headland which lies near the heading mark will seem to project further to seaward than it really does. Allow for this by moving the EBL just inland of the headland, by an angle equal to half the beamwidth.**

FIXING BY RADAR RANGES

Together, heading error and beamwidth error conspire to make radar bearings pretty unreliable. A far better bet is to use radar ranges. The boat's heading is irrelevant when measuring range, so the largest error is eliminated straight away. Moreover, most sets are inherently better at measuring range than bearing – range accuracy is typically within 1% of the range scale in use.

The chartwork involved may be unfamiliar to those accustomed to basic traditional methods, but it lends itself to work on a small chart table or a bouncing boat.

Again, the process starts with choosing suitable landmarks; it's essential that the features chosen are identifiable, and they should ideally be as close as possible.

➤ **To fix your position using radar ranges, use the VRM to find the ranges of three widely-spaced, recognizable contacts. Choose objects that are reasonably close to the boat, since this improves accuracy.**

Suppose we have a lightship 3.75 miles away. We must be the same distance from the lightship as it is from us, so our position lies somewhere on the circumference of a circle centred on the lightship and with a radius of 3.75 miles. A pair of compasses can be used to draw this circle on the chart, giving one of the position lines required to produce a fix. It's a curved position line, but that doesn't matter: it's just as valid as the straight line derived from a compass bearing.

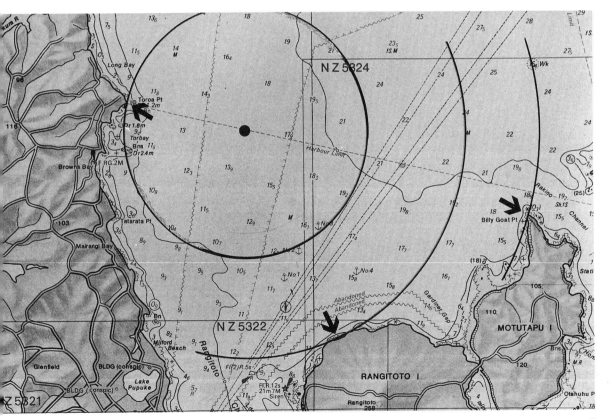

Using the VRM

The pattern of range rings provides an instant means of measuring range. All but the very cheapest sets, though, are equipped with a far more accurate system called a variable range marker, or VRM. As its name suggests, the VRM is an adjustable range ring whose distance from the centre of the screen can be varied by the operator. The radius of the VRM is indicated by a numeric display – usually in nautical miles – in one corner of the screen.

A VRM is far easier to build into an analogue display than an EBL, so all but the most primitive sets will have one. It's created by a completely different process than the VRM of a raster-scan display, though, so it's usually called a range strobe.

> Once you have all the ranges noted, take a pair of compasses and mark each range on the chart, with the compass point on the contact in each case. Short arcs are sufficient. The point where these circular position lines cross marks your position.

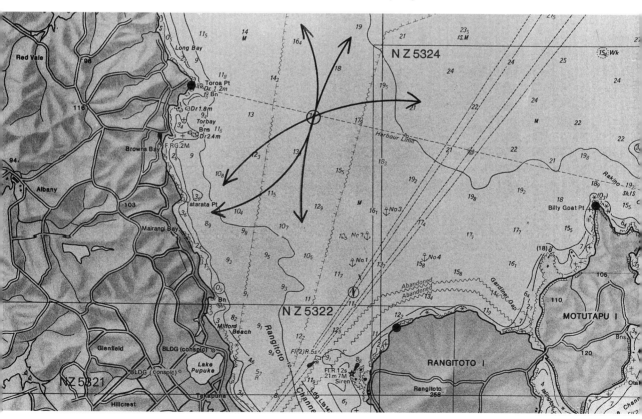

The tracker ball or keypad-operated cursors which are gradually taking over the role of EBLs are doing the same for the VRMs, although when measuring range the time savings they offer tend to be less pronounced than when measuring bearing.

Whichever system you have, the range is measured by moving the VRM, strobe, or cursor until it just touches, but does not cut, each of the chosen features in turn. To make the most of the system, as well as to make accurate positioning as easy as possible, it's good policy to use the shortest possible range scale for each measurement, even if

this means changing range part-way through a fix.

MIXING POSITION LINES

The radar ranges of three different landmarks will usually produce a good fix, the three circles intersecting to form a small triangle, or 'cocked hat'. Just as in traditional navigation, the size of the cocked hat gives a good indication of the reliability of the fix.

There are occasions, particularly when making a landfall, when it is not possible to

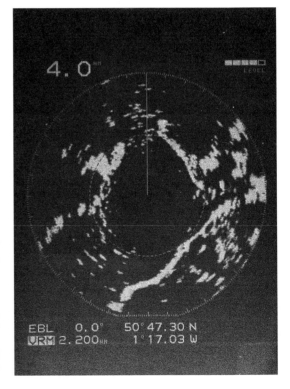

identify three suitable landmarks. But there's no rule that says all your position lines have to be of the same type. You could use a radar range and bearing of a single object to provide two position lines crossing at the perfect angle of 90°. Better still would be a visual bearing crossed with a radar range. Either way, you will need a third position line to give a reliable fix.

▶ **The sequence below shows a fix using radar ranges. From left to right, the VRM is aligned on Calshot Spit (2.14 miles), Hill Head (2.2 miles) and Old Castle Point (1.38 miles). The circular position lines are then scribed on the chart to give a crisp fix.**

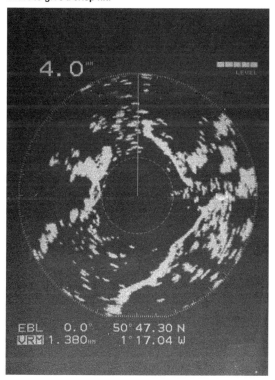

7. Collision Avoidance

Where radar most definitely comes into its own is in determining the risk of collision. For that, it's second only to the Mark One Eyeball – and in some respects even better. Not for nothing does Rule 7b of the International Regulations for Preventing Collisions at Sea state that 'Proper use shall be made of radar equipment if fitted and operational. . .'

There's more to making 'proper use' of it, though, than just casting an eye over the screen now and then. You certainly cannot drive a boat safely through thick visibility by steering to miss the bright blobs on the screen.

The mere presence of a contact does not mean that there's a danger of collision, any more than seeing another vessel means you're likely to hit it. What's important is your movement relative to each other. If you can see the other vessel, it's very simple to determine the risk of collision by taking bearings of it: if it gets closer without an appreciable change of bearing, there's a risk of collision, so one or the other of you will have to take avoidance action.

In practice, though, you might not need to whip out the hand-bearing compass for each

▶ You can assess the risk of collision visually by taking bearings of an approaching vessel using the hand-bearing compass. If the bearings stay the same, you are on a collision course.

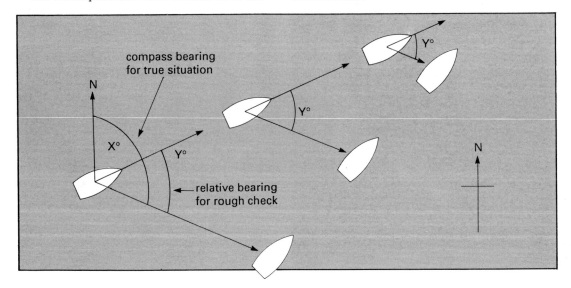

compass bearing
for true situation

N

X°

Y°

Y°

relative bearing
for rough check

N

and every vessel that heaves into view. You might very well start by making a rough check, lining the other vessel up with a guardrail stanchion or some other fixed part of your own boat. If, after a few minutes on a steady course, the other vessel appears to have moved away from the stanchion, then you don't need to do any more than keep an eye on it.

Evaluating collision risks by radar uses exactly the same principles.

The radar equivalent of the rough check is to move the EBL to cut the contact, as if to take a bearing of it. If after two or three minutes the contact appears to be sliding along the EBL towards the centre, its relative bearing has not significantly altered, so there must be some risk of collision.

◆ **With radar, you can use the EBL to run a rough check on another vessel. If your course is steady and the approaching contact stays on the EBL, you may well be heading for a collision.**

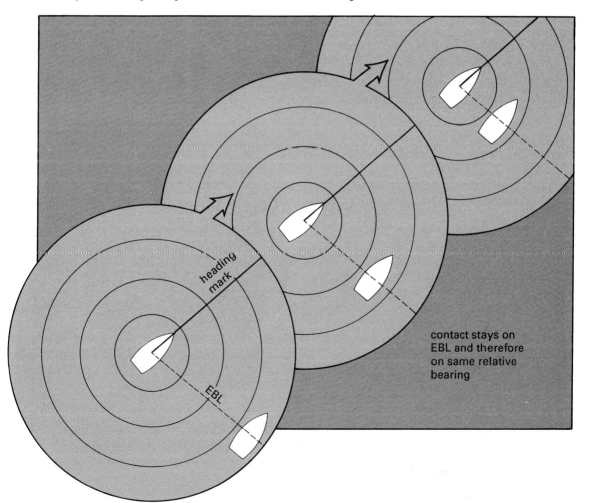

heading mark

EBL

contact stays on EBL and therefore on same relative bearing

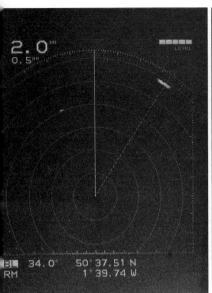

◆ **The sequence above reveals a contact approaching on a steady relative bearing of 34°. But the actual course, speed and aspect of the vessel are unknown.**

▶ **Plotting on screen using a grease pencil may help you make a better assessment of the situation.**

SIMPLE PLOTTING

Phase two, on radar, is called plotting. Like taking compass bearings and watching the other vessel through binoculars, it is intended to provide a more accurate and detailed assessment of the situation.

The easiest method is to plot directly on the screen, using a light-coloured Chinagraph (grease pencil) to mark the position of each contact on the face of the screen itself. This works well on a large screen, but many small boat radars are so

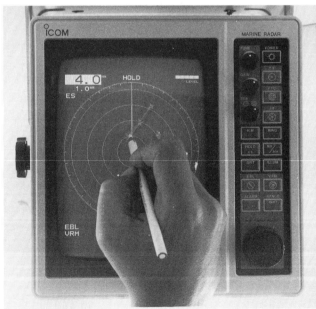

small that it can never be more than a rough and ready guide. Plotting on screen is still very helpful, though, as a means of helping keep track of several contacts at a time when operating in crowded waters.

A more accurate method is to mark the exact position of each contact on a plotting sheet – a paper representation of the radar screen. Pads of plotting sheets can be bought from good chandlers, or you could photocopy page 96, which is excluded from copyright for that purpose.

Whichever system you use, if you mark the positions of each contact at regular intervals – say every three minutes – you will quickly draw up a picture showing their movements across the screen. After a couple of plots, it's reasonably easy to predict each contact's likely future movement by continuing the line of its past progress. So if one contact appears to be heading straight for the centre, and nothing changes, sooner or later it will get there. In other words there will be a collision.

▶ **Plotting an approaching contact on a plotting sheet allows you to work out its course, speed and aspect. The first step is to mark each contact on the sheet at regular intervals. Here the contact appears to wander slightly because of variations in your own course, but it is clearly heading straight towards the centre of the screen and a possible collision.**

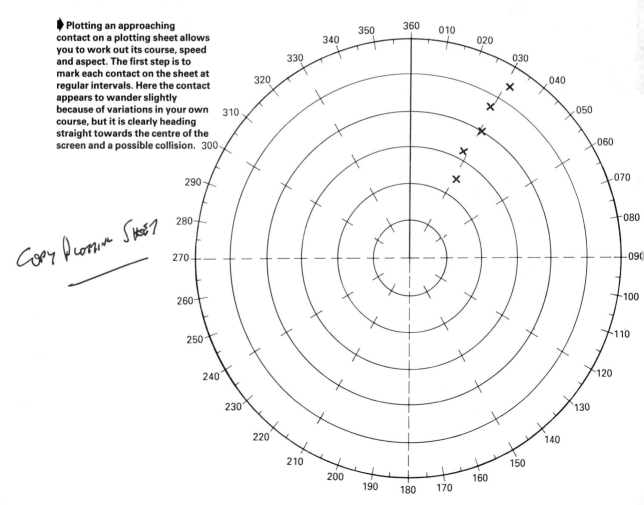

Copy Plotting Sheet

CLOSEST POINT OF APPROACH

If the line of movement passes close to the centre without actually going through it, then the chances of collision are less, but there is still a risk of an unpleasantly close encounter. You can find out just how close that 'closest point of approach' (CPA) is likely to be by measuring the distance from the centre of the screen to the closest part of the projected track.

By comparing the rate at which the contact is moving along its plotted track with the distance it has to go before it reaches its CPA it is even possible to estimate the time remaining before it gets there.

While you're doing all this, it's worth bearing in mind that the man on the bridge of the other vessel is likely to be doing the same thing, and that if it's a large ship in open water he's likely to start feeling nervous if his calculations show a CPA of less than two miles. That may well seem excessive to a small boat skipper, but in dealing with ships it does no harm at all to adapt your thinking to their scale, and to regard anything within two miles as a 'near miss'.

◆ **Drawing a line through the past positions of a contact shows whether it will pass ahead or astern, and its closest point of approach (CPA). In this case the target will probably pass 0.6 miles ahead and its CPA will be 0.5 miles on the starboard bow.**

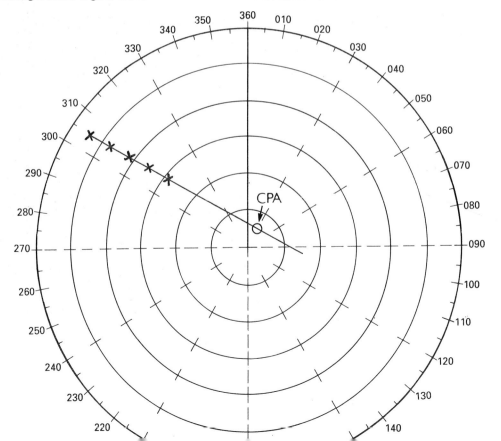

COURSE AND SPEED

To decide who should stand on, who should give way, and what avoiding action is required, you need to know something about the other vessel's course and speed. For this, the simple plotting used to assess the risk of collision isn't enough.

The diagram below illustrates the problem. It shows a series of plots, taken at three-minute intervals, of a small contact which is closing from the port bow. In 12 minutes its range has decreased from 5½ miles to 2½ and its CPA is likely to be within ¼ mile. This is too close for comfort, so what should we do about it?

As the range of the contact is decreasing at the rate of one mile every four minutes, and the last plot put him 2½ miles away, we have less than 10 minutes in which to work out what is happening and decide on a suitable course of action.

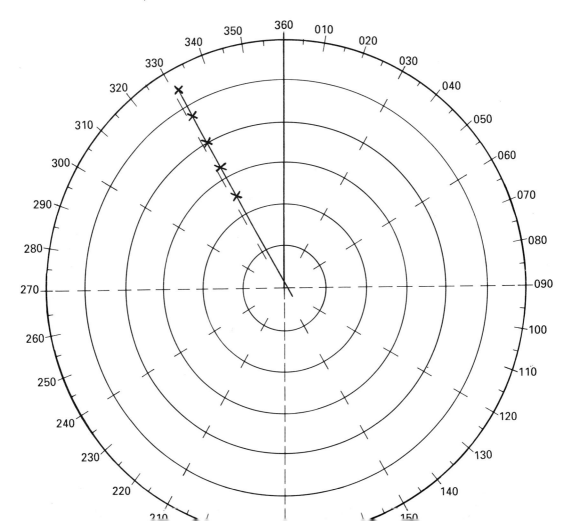

Example 1

Assume, for the moment, that we're in a sports cruiser, doing 25 knots, and imagine that the other vessel had dropped some kind of marker at the moment of our first plot.

Because we are heading straight up the screen, but remaining at its centre, the contact of any stationary object will move straight down the screen at a speed equivalent to our own. So a plot of the marker would look like the diagram below.

◀ **Building on the plot opposite, an imaginary marker is plotted from the position of the first contact, moving down the screen at a speed equivalent to your own (25 knots). A line drawn from the fifth marker to the fifth contact gives the vessel's true course and speed.**

The other vessel's movement through the water is shown by its position relative to the marker: in this case, it has moved 2¾ miles from the marker in 12 minutes, so its speed is 2¾ × 5 = 13¾ knots. More significantly, its course is represented by the direction of the line from the marker to the contact, which is converging with our own heading at an angle of about 30°.

So the plot shows that we are catching up with a slower-moving vessel, whose course is converging with our own in such a way that we are approaching his starboard quarter. In other words, we are an overtaking vessel within the terms of the Collision Regulations, so we are required to keep clear.

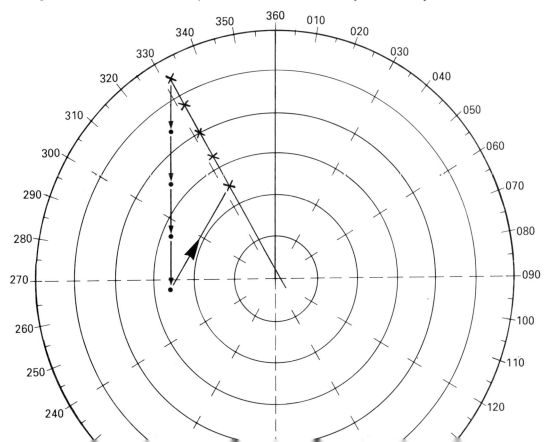

Example 2

Now let's look at an identical sequence of plots, but this time taken on a yacht, making five knots under engine.

The imaginary marker would still move straight down the screen, but now, because our own speed is so much lower, it too would be moving more slowly, covering just one mile in the 12 minutes between plots 1 and 5.

The geometry has changed dramatically, but the principle is the same: the distance the other vessel has travelled in 12 minutes is represented by the distance between its last plotted position and the imaginary marker, and its course, relative to our own, by the angle at which the line drawn from the marker to the last plot cuts the heading mark.

So this time we are looking at a vessel doing 11 knots on a course at 135° to our own, in such a way that we are looking at her starboard bow. The whole situation is different: now the Rules require the other vessel to alter course, and us to hold our course and speed.

➤ **In this case the basic plot is the same as the last, but because your own speed is only five knots the 'marker' does not travel so far. Drawing in the course/speed line reveals a very different situation.**

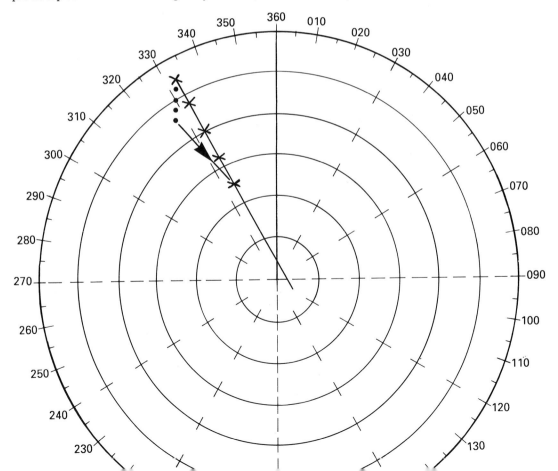

It is essential that you understand your obligations under the Collision Regulations, so that you know what to do in any close-quarters situation. *The Small Boat Guide to the Rules of the Road,* also published by Fernhurst Books, will tell you all you need to know.

TO SUM UP

It would have been virtually impossible to decide exactly what was going on in either of these cases without plotting. It's cases like these that explain why Rule 7b, having said 'Proper use shall be made of radar equipment . . .' goes on to add '. . . including long range scanning to obtain early warning of risk of collision, and radar plotting or equivalent systematic observation of detected objects'.

One occasionally hears reports of 'radar assisted collisions', which just goes to show that even professionals sometimes misinterpret their radars. But the plotting routine is simple, especially if you keep in mind the idea of an imaginary marker buoy:

1 Use the EBL as a rough guide to establish whether there is a risk of collision.

2 Plot any doubtful contacts at regular intervals, either on screen or on a separate plotting sheet.

3 Project the line of plots onward, beyond the centre of the screen, to establish the contact's CPA and the time remaining before it gets there.

4 From the first plot, draw a line straight downwards to represent the movement of an imaginary buoy.

5 Mark the track of the 'buoy' at a point corresponding to the distance you would have moved in the time between the first and last plots of the real contact.

6 Draw a line between the marked position and the last plot to represent the actual course and speed of the target.

8. Blind Pilotage

The other aspect of navigation at which radar excels is pilotage – conducting a vessel along a safe pre-planned track, in waters so confined that conventional fixing and chartwork are impractical.

Not only is its inherent accuracy better than any other electronic navigation aid – with the possible exception of the GPS satellite system* – but its graphic presentation, showing the boat's position in the context of its immediate surroundings, can be interpreted almost instantaneously, without having to apply corrections or get involved with chartwork.

But it's important to keep in mind that radar is not a magic lantern. It's easy to be seduced by the technology and to forget that boats – even boats with radar – are subject to the effects of leeway and drift. The key to success is meticulous planning: working out well in advance what should happen, what might happen, and what you can do to make life as simple as possible. This will invariably mean making pilotage a two-person job, with one on the radar and one on the helm, and it may mean following a different track to the one you'd use if you were piloting by eye.

---

*The Global Positioning System of satellite navigation is not yet (1990) fully operational, but is usually quoted as having errors of less than 25 metres in 95% of fixes achieved. It is possible, however, that the American Department of Defense, who own and operate the system, will degrade the accuracy available to civilian users to about 100-200 metres once all the satellites are operational.

RAPID FIXING

Rapid fixing doesn't really qualify as a pilotage technique at all, because it's not a method of conducting a vessel along a pre-planned track. Nevertheless, it can be very useful in waters where there are so many hazards that conventional fixing would be unacceptably slow.

It involves pre-planning a series of fixes, each made up of two radar range lines, and drawing one of each pair of position lines on the chart in advance. In practice, this means picking a series of easily-identified, radar-conspicuous landmarks – ideally ones lying fairly close to the intended track. Next, having decided on the range scale you are going to use, draw a series of concentric rings around each of your key landmarks, at intervals corresponding to the distance between the range rings on your chosen range scale.

Then, whenever a key landmark crosses a range ring on the screen, you know at once that you are somewhere on the corresponding ring drawn on the chart, and need only measure and plot the range of one other object in order to complete the fix.

▶ **This chart shows range rings drawn around Lion Island, Box Head and Barrenjoey Head. The ranges correspond to the range rings on the radar screen, so if, say, Box Head crosses the one-mile ring on the screen, you know you are somewhere on line A.**

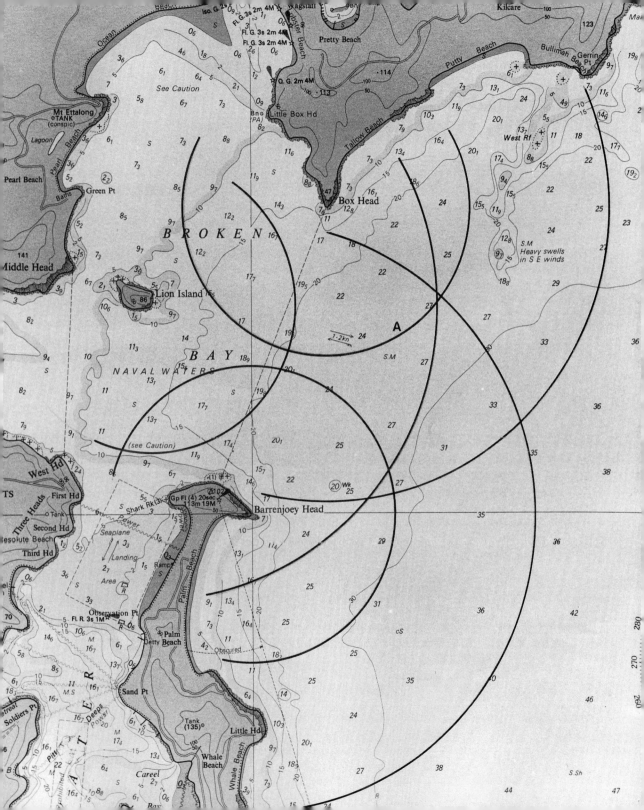

DRIVING THE SCREEN AROUND

'Driving the screen around' really needs very little explanation, as it's the sort of approach most radar operators tend to use almost instinctively.

It's rather like those computer video games, in which you are assumed to be in control of some kind of vehicle. The car, spaceship, or whatever it may be doesn't necessarily appear in the picture, but even if it does, it remains stationary in the centre of the screen. The illusion of movement is created by making the 'scenery' move past it.

This, of course, is exactly what happens in the picture on a head-up, relative-motion radar, so the techniques you'd use in a video game can be applied directly to pilotage.

Suppose, for example, you want to pass between two buoys that you have identified on the screen as a pair of small contacts 30° off the starboard bow. You don't need a book to tell you that altering course 30° to starboard will leave you pointing straight towards them. As you alter course, the whole picture will rotate so that when the boat is pointing straight at the gap, the heading mark passes between the two contacts. This means that you can find your way along a well-buoyed channel by keeping the heading mark pointing straight along the line of the channel.

➤ **The diagram on the left shows the radar picture before altering course up a buoyed channel. As you change course (below) the picture rotates on the screen, and you straighten up when the heading mark is pointing directly between the buoys.**

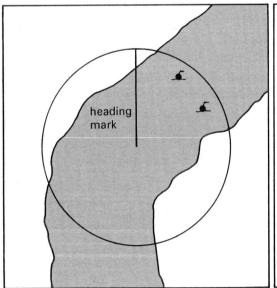
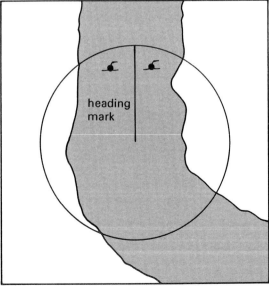

Channels which are indicated only by shore marks, such as leading lights, aren't quite so simple. The marks just don't show up on radar!

All is not lost, though. It would be most unusual not to have some radar-conspicuous object, such as a headland, nearby. If you say to yourself 'I need to keep between half and three-quarters of a mile off the headland', it's not too difficult to keep the heading mark that distance off the headland. The job is even easier if you draw a Chinagraph line across the screen, the right distance away from the heading mark and parallel to it.

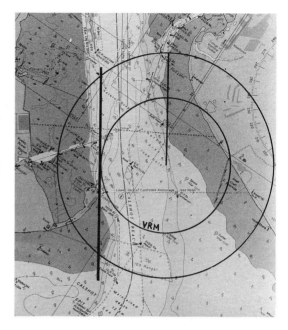

➤ If there are no conspicuous buoys in the channel, you can draw an offset heading mark on the screen and keep it aligned with the headland. The chart and screen on the right show an offset heading mark in practice, aligned with Calshot Spit.

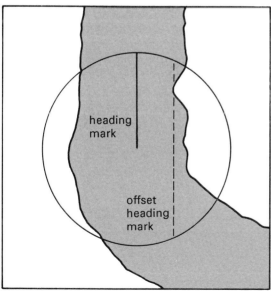

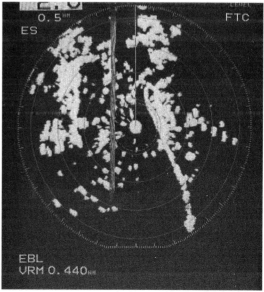

The danger with this 'seat of the pants' method is that in a strong tidal stream or cross-wind, your track (over the ground) may not correspond to your heading. As the wind or tide push the boat sideways, the contact you're aiming at will appear to slide away from the heading mark. Even if you alter course to bring it ahead again, you will still be off track and still sliding sideways. The end result is that your track won't be a straight line towards the mark, but a curve. The curve may take you into shallow water,

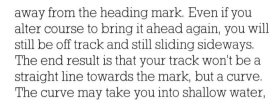

➤ **If you simply aim at a buoy by keeping it on the heading mark the tide may push you sideways into danger before you realize it.**

➤ **The same thing can happen if you align an offset heading mark on a single landmark. As you alter course to keep the mark in place you drift off track.**

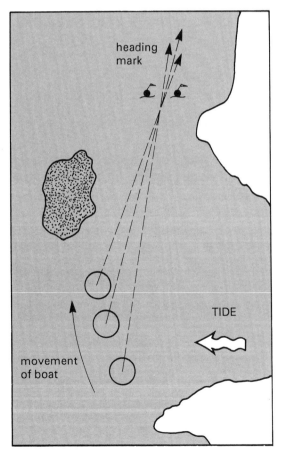

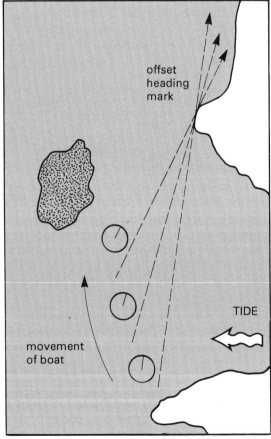

or it may disorientate you: either way it's something to watch out for.

OFFSET LEADING LINES

In conventional visual pilotage, you could overcome this problem by keeping some object beyond the next buoy in transit with it. Radar's poor bearing discrimination – not to mention its complete inability to distinguish small objects ashore – usually makes this impractical. The corresponding radar technique is to use a variation of what some big-ship navigators know as an offset leading line.

This involves picking your intended track so that it passes two radar-conspicuous objects at equal distances. In other words, the track has to be parallel to the straight line between two landmarks.

That straight datum line can be reproduced on the radar screen with the straight edge of a scrap of paper or a postcard, held against the screen so that it cuts through both contacts. The boat's distance off the datum line can be measured by setting the VRM so that it just touches the edge of the card. Conversely, if you set the VRM to the required distance from the datum line, you can see at a glance which way you need to alter course in order to stay on track.

▶ **Using an offset leading line (or offset transit) aligned with two landmarks overcomes the drift problem. Having decided on your leading line you simply set the VRM to the required distance off, and keep it just touching the line as you head up the track.**

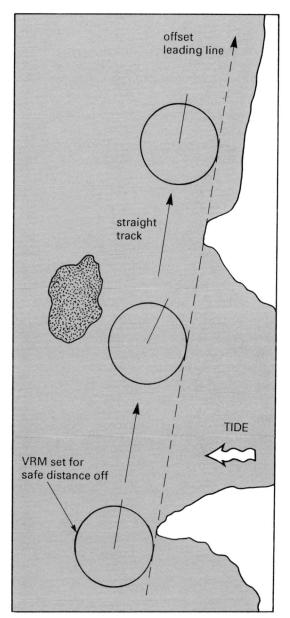

A further variation on the offset leading line can be used to determine the right moment to alter course. Suppose, for example, that we're on our way up the Solent, heading for the Hamble River. The line up Southampton Water is about 330°, for which we might well use an offset transit – keeping the line between the tip of Calshot Spit and the southern end of Fawley jetty 4½ cables (0.45 mile) to port. Approaching from the west, therefore, we have to go 0.45 mile past the line before turning to port. Again, the line is best represented on the screen by a hand-held strip of paper, and the offset measured with the VRM.

This may seem a complicated way of doing things, especially as this particular channel is well-buoyed. But bear in mind that radar can't distinguish between one buoy and another, or between a buoy and a boat. Using large, easily-identifiable objects like the spit and the jetty removes the risk of confusion.

◂ Held up to the screen to align with two identifiable landmarks, the edge of a Douglas Protractor provides an excellent offset leading line. Here it is aligned with the tip of Calshot Spit and the southern end of Fawley Jetty, as shown on the chart opposite. Setting the VRM to 0.45 miles allows you to judge the moment to turn to port, as well as the correct route up the channel.

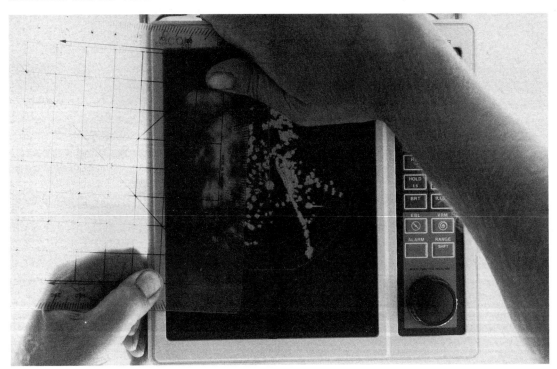

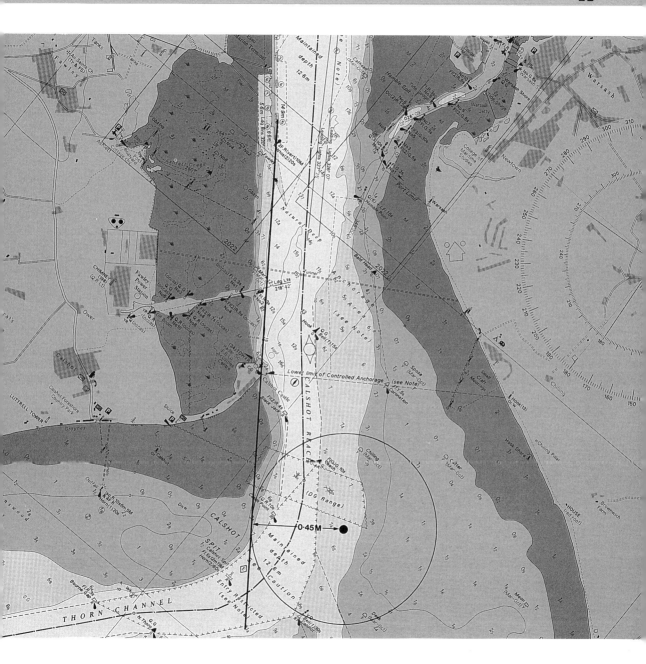

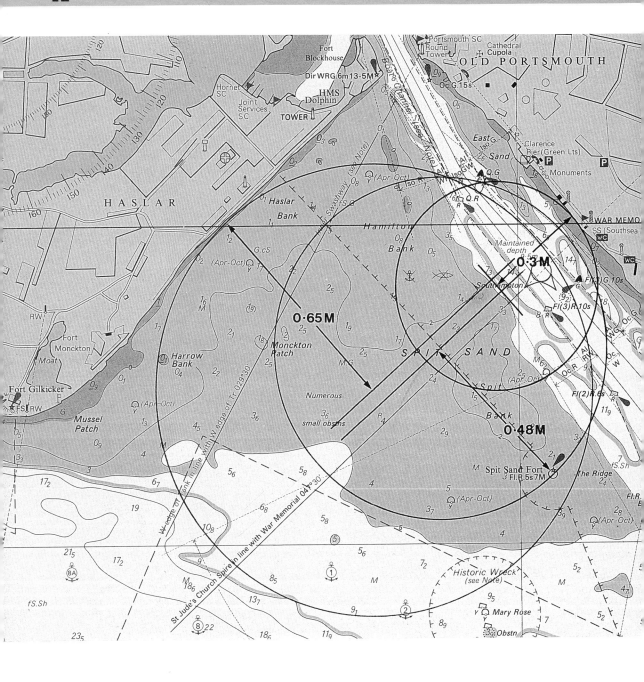

Radar can also provide clearing lines. In visual pilotage, a clearing line is usually based on a compass bearing of a landmark. Approaching Portsmouth, for example, you might well use clearing bearings to pick your way through the swashway between Spit Sand and the Hamilton Bank: so long as the bearing of the War Memorial is between, say, 050°T and 060°T, the boat must be somewhere in the safe channel.

Using radar, the principle of clearing lines is even easier. The Hamilton Bank is within 0.65 mile of the shoreline, so if the shore is outside the VRM at 0.65 mile, the boat must be clear to the south of it.

In this particular instance, of course, there's a second hazard to avoid: Spit Sand lies just to the south.

So you'd almost certainly want to use two clearing lines. The first, on the left-hand side of the screen at 0.65 mile to guard against Hamilton Bank; another, on the right-hand side at 0.48 miles to clear Spit Sand, using Spit Sand Fort as its reference, and possibly even a third, ahead and at 0.3 mile, to stop you turning too soon. In cases like this, where there are several clearing ranges to remember, it can help to draw them onto the screen in Chinagraph. Just don't change range scales in the middle!

◀ **The approaches to Portsmouth, showing the route through the swashway, the hazards to north and south and the clearing lines (see text above).**

▶ **Setting the VRM to give you a safe distance offshore will enable you to search for a harbour mark with confidence. As long as you keep the shoreline outside the range ring you are in no danger of grounding the boat, so you can keep going until you spot the buoy.**

Radar clearing lines can be very useful even in perfect visibility, especially for that nail-biting section of a passage when you've found the harbour you're heading for, but can't see the marks which will guide you in. Once you've decided that there are no hazards more than, say, a mile offshore, you know you can safely go inshore until the shoreline nudges the one-mile range ring. This kind of clearing line is difficult to draw on the chart with any accuracy, especially if the coastline is uneven but without clearly-defined headlands. Nevertheless, it's perfectly valid.

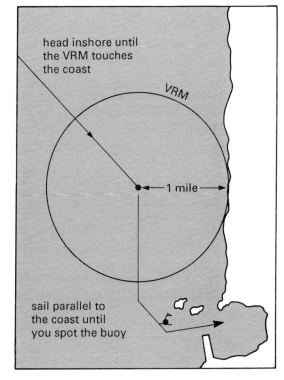

head inshore until the VRM touches the coast

VRM

◀—1 mile—▶

sail parallel to the coast until you spot the buoy

9. North-up Radar

True in MAGNETIC?

North-up displays are a relatively recent phenomenon in small-craft radars, but are potentially one of the most significant since the development of the PPI – even including the advent of raster-scan.

A north-up display is linked to an electronic compass and, as the name suggests, its picture is rotated so that north is at the top. The most obvious effect – that the picture corresponds more closely with the chart – is relatively insignificant, and in some respects a drawback. Far more important is the fact that it retains the same orientation regardless of accidental or deliberate changes of heading. In doing so, it removes the single biggest source of bearing error.

Even so, changing from head-up to north-up has relatively few advantages in collision avoidance. It may make plotting more accurate and slightly easier, but at the same time it makes comparing what you can see out of the wheelhouse windows with what you see on the radar much more difficult: if you happen to be heading south, anything on the right-hand side of the screen will be on the port side and vice versa.

It certainly does not remove the need for careful, regular plotting.

With a head-up display (left) a contact swings about as the boat yaws, but the display gives an accurate idea of its position off the port bow. A north-up display (below) is more stable, but less easy to understand.

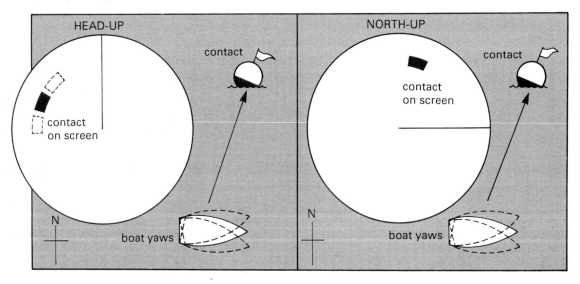

In the context of navigation, the biggest virtue of north-up is not the orientation – again something of a two-edged sword – but stability, and the improved bearing accuracy that comes with it. A second fruitful source of errors is removed because north-up does away with the need to convert from relative bearings, although you'll still need to allow for variation and deviation.

PILOTAGE

It's in pilotage that north-up really proves its worth. Orientation now becomes immensely significant, because it makes it possible to work out how landmarks will move around the screen as the boat moves along her intended track.

You could, in theory, choose a single distinctive feature, work out what its range and bearing should be at every stage of the pilotage passage, and plot its predicted movement on the screen with a Chinagraph pencil. The result, incidentally, would be the same shape as the intended track drawn on the chart, but rotated through 180°. Then, as the boat moves, the contact of the landmark should follow the Chinagraph line: if it doesn't, then the boat must be off track, and you could alter course to correct it.

In practice, though, radar's inherent lack of accuracy and discrimination in bearing is a limitation; besides, it is much easier to break even a short stretch of pilotage into separate stages.

Keeping track
The most fundamental task in pilotage is to maintain a pre-planned track. This may well involve altering course to counteract the changing effects of wind and tide. That's

why, with a head-up display, it's advisable to use an offset leading line.

With a north-up display the leading line becomes far less significant. If you wanted to go straight towards a buoy, you could set the EBL to the intended track, and then adjust your course so that the contact of the buoy made its way straight down the EBL towards the centre of the screen. Because the display is stabilized to the compass, this is simply another way of steering on a line of bearing towards the buoy.

In real life, you'd be more likely to want to pass a buoy or landmark at a specific distance. Given the choice, in fact, it's more accurate to use a feature that's off to one side of your intended track. In that case, you'd set the EBL to the intended track and draw a line parallel to it, the appropriate distance to one side.

So if your intended track is 085°C and it leaves an island 600 yards to port, you'd set the EBL onto 085° and draw a line 600 yards to the left of it – using the VRM to measure the distance.

So long as the boat is on track, the radar image of the island will then move along the Chinagraph line. If it slides to starboard of the line, then it means you're being set to port and should alter course to starboard – and vice versa.

You may not need to be that pernickety about distances. If it doesn't really matter whether you're exactly 600 yards off the island so long as you're between ¼ mile and ½ mile, it's worth drawing two lines, one at ¼ mile and one at ½ mile, corresponding to the two sides of the 'corridor'.

▶ **With a north-up display, you can use a single landmark such as an island to stay on track. Draw a guide line on the screen to represent the correct distance off, and keep the edge of the VRM on the line. As long as you stay on track the landmark will slide along the line. The photograph below shows the scene, with the camera pointing north.**

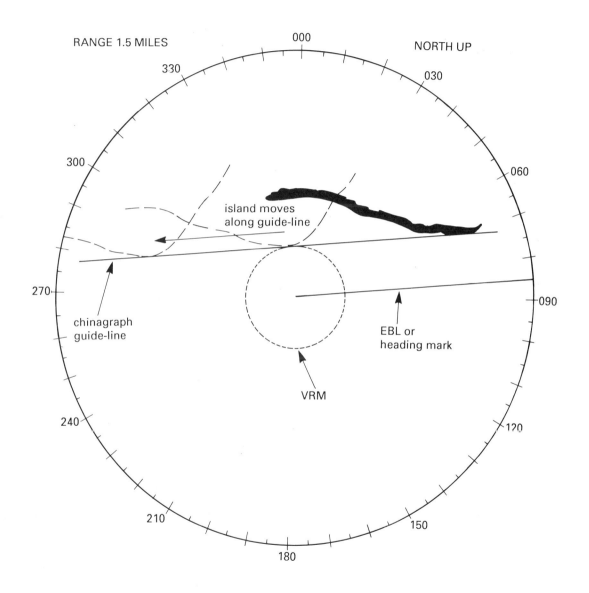

RANGE 1.5 MILES

NORTH UP

island moves
along guide-line

chinagraph
guide-line

EBL or
heading mark

VRM

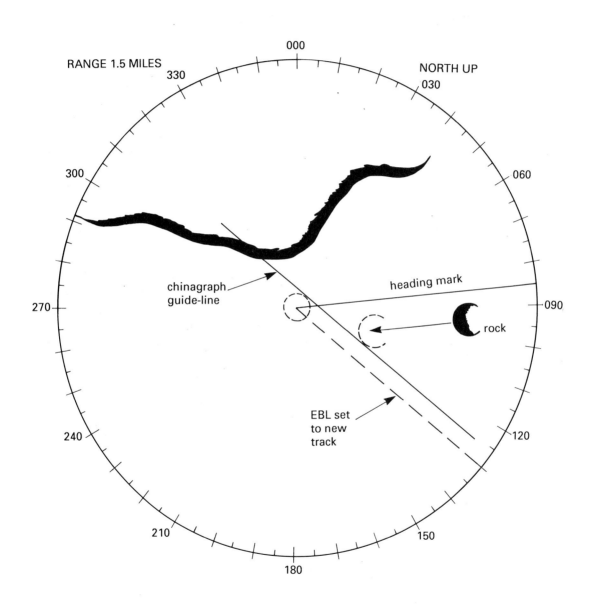

RANGE 1.5 MILES

NORTH UP

chinagraph
guide-line

heading mark

rock

EBL set
to new
track

ALTERING COURSE

A very similar principle can be used to plan an alteration of course so that, at the end of the turn, you end up on the right track.

Suppose your old track is 085° and you're intending to alter course to starboard onto a new track of 130°, so as to leave a conspicuous rock 200 yards to port.

Because the north-up display is compass-stabilized, there's no reason why you shouldn't draw the guide line onto the screen well in advance by setting the EBL to 130° – the new track – and drawing a line parallel to it and 200 yards to port.

The movement of the boat will make the contact of the rock move across the screen towards the line. Just before it touches is the moment to alter course.

The key to success lies in choosing the right landmark. It's no good choosing one that will be ahead or astern on the new course, because that would be dependent on the radar's ability to measure bearing – and even with a north-up display, that's not its strong point. The best mark to choose is one that will be as nearly as possible abeam of the new track at the moment you alter.

CHANGING RANGE

Pilotage is most often used at the beginning and end of a passage, to get out of one harbour and into another. In other words, it leads you from the open sea into confined waters, or vice versa. Almost by definition, then, it will involve changing the range scale of the radar. This calls for some care at the planning stage, because you don't want to be rubbing out and redrawing Chinagraph lines on the screen every time you change the range scale.

So it's essential to make sure that the guide lines are plotted on the screen using the same range scale as you will be using when you come to put it all into practice. Then make sure that you really do use the range scale you had thought you would. Memory isn't enough: it's far better to use a notebook, or even to mark planned changes of range scale on the chart. As a double check, and if the size of the screen permits, you could even make a note of the range scale alongside each Chinagraph line, or number the lines so that you can quickly and easily refer back to your notes. Time is often short in pilotage.

◀ Using north-up radar to pick the right moment to alter course. The guide line is drawn on the screen in advance, at a safe distance; the contact of the rock moves across the screen parallel to the heading mark, and you alter course just before it reaches the guide line.

10. Special Features

North-up is such a significant feature that it warranted a chapter of its own. But the kind of technology which made raster-scan and north-up possible has given manufacturers the opportunity to include a host of other refinements.

Specifications vary widely, depending on the manufacturer, on how long a particular model has been in production, and on its position within the model range. Undoubtedly, as time goes by, some features which are rare at the moment will become commonplace and others will take over as the luxuries which distinguish a top-of-the-range model from its lesser siblings. At the moment, though, you might well come across the following:

- Automatic tuning
- Colour display
- Course-up display
- Cursor
- Echo plotting
- Floating EBL
- Floating VRM
- Guard zone
- Heading mark deletion
- Interfacing
- Manual pulse length selection
- Multiple VRMS
- Multiple EBLs
- Off-centring
- True motion display

The names given to these 'bells and whistles' vary from one manufacturer to another, as do the details of their precise effects and method of operation. They all have their uses, but that doesn't mean that a radar with all the goodies is necessarily better than one without. Added features inevitably mean increased complexity – and if your radar is so complicated that you can't operate it without the instruction manual open in front of you, you're not likely to get the best out of it.

Automatic tuning

The implication of automatic tuning is obvious – it saves the operator having to fine-tune the radar himself. It's quite useful when you first switch on, but does not work unless there is a suitable target within range and in any case is seldom as effective as manual tuning.

Colour display

Colour displays are becoming more widespread, particularly in pleasure craft. Most colour displays offer several different formats, giving one, two, three or four different colours, on a choice of one or two different backgrounds.

The significance of the colours varies: some use different colours to indicate different signal strengths, others use colour purely to differentiate between genuine contacts and things such as EBLs and VRMs. It's a popular misconception about colour radars that they show land in one colour, ships in another colour and buoys in yet another. They can't – though it is true to say that because land generally produces a strong echo it will usually appear in a 'stronger' colour than anything else on the screen.

Perhaps the most significant virtue of a colour display is that it allows you to change the background colour. In daylight, for example, you might choose a blue background with red and yellow contacts, because that combination is very clear and easy to read even in bright sunlight. At night, however – even with the brilliance turned right down – a big glowing screen wreaks havoc with your night sight, so it is useful to be able to switch to a black background.

The main disadvantage of colour is that by drawing attention to strong contacts, colour may distract you from weaker but more significant contacts.

Course-up display

Course-up displays have something in common with north-up, in that the picture is stabilized by linking the radar to an electronic compass. Instead of putting north at the top, however, it puts the course, as set on the boat's autopilot, upwards.

At first sight the result is indistinguishable from a head-up picture, but with the important difference that the picture does not swing from side to side as the boat yaws – so it's ideal when the radar is being used for collision avoidance.

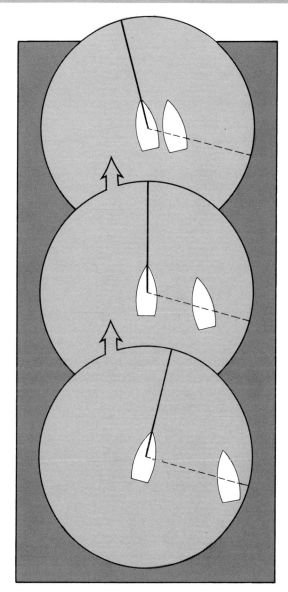

⬆ A course-up display is stabilized to the autopilot, so a contact does not swing about as the boat yaws. This means that a contact on a collision course will stay on the EBL regardless of your actual heading.

Cursor

An electronic cursor, usually in the form of a hairline cross or a small square, can be brought into the picture on some of the latest radars. It can be moved around the screen using either a panel of four control keys or a tracker ball, its range and bearing from the centre being displayed on the screen.

Echo plotting

The term 'Echo plotting' or simply 'Plot' usually refers to a facility which mimics the faint trail left behind each contact on an analogue display. The trail, showing the past position and therefore the movement of each contact, is useful as a simple initial check on a potential collision situation.

The length of trail varies, and on some sets it can be selected by the operator, but few small-craft sets offer an echo plotting facility with a memory long enough to relieve the operator of the need to plot potential collision situations manually.

Floating EBL and VRM

Floating EBLs and VRMs are also used in conjunction with a cursor: they measure the range and bearing from the cursor to any other point on the screen. With a bit of practice they can also be used in simple pilotage, as an alternative to drawing Chinagraph lines on the face of the screen.

Guard zone

All but the most basic sets now incorporate a guard zone facility, which sounds an alarm whenever a contact enters a predetermined area of the screen. At the most basic level,

this may simply mean that the alarm sounds whenever there is a contact inside a chosen range, but to prevent sea clutter giving rise to too many false alarms, most guard zones also have an inner limit, or take the form of a ring rather than a circle.

More sophisticated variants enable the operator to reduce the guard zone to a semicircle or a quarter circle, or even to define its left-hand and right-hand edges completely at will. This is particularly useful if you are running close inshore but want to use the guard zone facility to give warning of contacts ahead, astern, or to seaward.

Here the guard zone has been set up to react to any contact approaching from the starboard bow – a vessel that you would have to give way to.

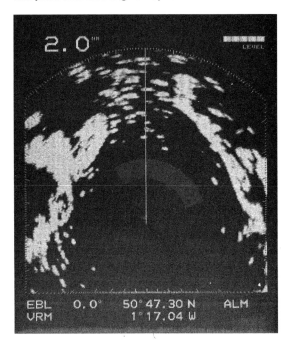

Heading mark and range ring deletion

The heading mark and range rings, (and, on north-up sets, the north mark) can obscure small contacts. Most sets allow you to delete the range rings, and many also allow you to remove the heading mark. Unlike the range rings, though, the heading mark control is usually self-cancelling, so that the heading mark reappears as soon as the button is released. It is good practice to turn the range rings off when they are not in use, and to delete the heading mark once every few minutes to check that there is nothing directly ahead.

➡ **Deleting the range rings makes the display less confusing, and it is easier to use the VRM. Turn them on only when you need them.**

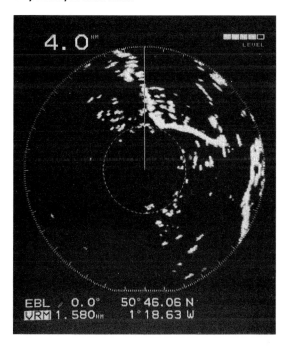

Interfacing

Many of the more advanced features, including north-up, require the radar to receive information from other navigation equipment such as a navigator or autopilot. This communication between instruments is known as interfacing, and is carried on in a code made up of very rapid electrical pulses sent through two-core cable.

Of course, it's essential that the radar understands the code being used by each piece of equipment to which it is connected. Fortunately, most marine electronic manufacturers have adopted a standard code defined by the National Marine Electronics Association (of America). As time has gone by, the original standard has been developed to handle more information, so you could now come across any of several variations of the NMEA standard. These are known as 0180S, 0180C, 0181, 0182, 0183, and 0183B. The latest, 0183 and 0183B, are the most comprehensive and therefore the most appropriate for radars. It's worth checking the NMEA standard on new equipment to ensure it will interface with your radar.

Manual pulse length selection

Pulse length has a significant effect on the performance of the radar, but is normally selected automatically when you select the range scale. A few sets, however, allow manual selection.

Choosing a longer pulse length increases the chances of the radar detecting targets that would otherwise produce impossibly weak echoes. The penalty is that range discrimination suffers dramatically.

Multiple EBLs and VRMs

On the face of it, there seems very little point in having more than one EBL or VRM: you can only set up and measure from one at a time anyway. Nevertheless, some sets have duplicate VRMs and EBLs which can prove useful if you want to do a quick check on two potential collision situations at once. They can also be used to mark a particular spot on the screen – as you might in pilotage, for instance.

Off-centring

The high speeds which can be reached by many small craft means that there is often only a very small risk of anything catching up with you from astern. It also means that the closing speed of any contact astern is unlikely to be high. On the other hand, it is very useful to be able to 'see' clearly for

some considerable distance ahead. Moving the centre of the picture downwards expands the effective range forwards at the expense of range astern: on a four-mile range scale, for example, off-centring would enable you to see six miles ahead but only two astern.

For those in sailing boats or low-speed motor cruisers, this facility should be used with caution: your low speed means that other vessels can catch up with you very quickly, and although they are obliged to keep clear of you, that won't be much consolation if you are unlucky enough to get run down.

Other, still more sophisticated sets have a similar facility, except that the centre can be moved in any direction. This can be very useful for navigation but, like off-centring, it should be used with some caution if you are relying on the radar for collision avoidance as well.

True motion display

True motion displays, in which the centre of the picture moves across the screen in step with the actual movement of the boat, are preferred by many professional navigators. That's because they give a clearer picture of what is going on around you: stationary objects, for instance, appear stationary on the screen. They are, perhaps, less appropriate for amateurs, because the plotting required to work out a closest point of approach (CPA) is significantly more complicated.

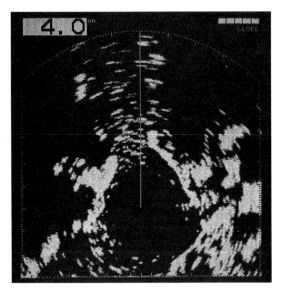

◀ **An off-centred display gives extra range ahead, but should be used with caution on slow vessels that may be caught up from behind.**

11. Installation and Maintenance

Not very many years ago, the idea of including a chapter on installation and maintenance in a book for amateur radar operators would have been as incongruous as a chapter on do-it-yourself brain surgery in a first-aid manual.

But times have changed: some sets, especially at the lower end of the market, can be carried out of the chandlery in a box and installed by anyone who can wire up a household plug. More sophisticated radars, however, may require internal adjustments that can only be carried out by a trained electronics engineer. Interfacing, in particular, is seldom as straightforward as it might seem. A safe and simple rule is to assume that if you have to take the back off, call in an expert. *

> POTENTIALLY LETHAL VOLTAGES PERSIST IN A RADAR SET EVEN AFTER IT HAS BEEN SWITCHED OFF OR DIS-CONNECTED.

But you can't leave the whole job to the installation engineer. A radar set is a heavy, bulky, and conspicuous piece of equipment,

---

*In the USA, a radar installation has to be approved by a Federal Communications Commission technician unless it is certified as having been pre-tuned by the dealer or manufacturer.

US regulations also require the ship's radio licence to be updated to include transmissions in the 9400MHz band.

so it's worth thinking carefully about where it is to go.

SCANNER LOCATION

At first sight it seems that the location for the scanner unit is obvious: as high as possible, to give the best possible range. Theoretically, that is certainly so, but in practice you may have to come down from that ideal.

To begin with, there's the matter of its weight. Putting something which may well weigh 25kg (half a hundredweight) at the top of a 12-metre (40ft) mast would put an enormous load on the mast and rigging, and could have the same effect on the boat's stability as removing half a ton of ballast!

Access is another important point, which limits the ideal height of the radar scanner to the length of a typical boatyard ladder – about 3–4.5 metres (10–15 feet) above the deck.

The lower limit is determined largely by safety considerations, because although radar manufacturers assure us that the power radiated by a small-craft radar is too small to cause any ill effects, there is no point in tempting fate by mounting it at eye level. Besides, there is the far more mundane danger of bumping your head on anything lower than about two metres (seven feet).

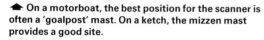

⬆ **On a motorboat, the best position for the scanner is often a 'goalpost' mast. On a ketch, the mizzen mast provides a good site.**

Sailing boats have the problem of sails and rigging to contend with – they somehow have to be prevented from fouling the scanner. Motor boat owners, relieved of that problem, may need to think about the radar's effect on the flybridge steering compass. 'Compass safe distances' are usually quoted by the radar manufacturer, but are rarely less than about a metre (three feet).

Finally, there's the matter of blind arcs to think about – not that they are much of a problem on small craft.

Taking all these considerations into account usually limits the available options, and on motor boats the most common location is on the wheelhouse roof or on a 'goalpost' mast.

Among sailing boats, ketches take most kindly to a radar scanner mounted near the mizzen spreaders. Sloops have more of a problem, because a scanner in the obvious place – just in front of the lower main spreaders – is likely to be fouled by the leech of the genoa. One of the best solutions, though it's seldom aesthetically appealing, is a stub mast about 2·5 metres (eight feet) high mounted near the transom.

⬆ **Mounting a scanner on a sloop can be tricky. Here a stub mast at the stern provides an effective solution, although it looks a little odd.**

DISPLAY LOCATION

The first requirement for the display unit is that it should be visible and within easy reach of the helmsman and/or navigator. The ideal spot, however, is usually already occupied by the compass – and as the 'compass safe distance' of a display unit is usually at least 30 cm (12 in) some compromise is inevitable.

In a sailing boat, a display mounted in the cockpit would be especially vulnerable to water or physical damage, so it's generally best to go for a position near the chart table. If possible, though, it should be arranged so that watchkeepers can keep an eye on it

through an open companionway or window, without having to go below.

In a motor boat the most obvious position is on or in the instrument panel, but this can obscure the helmsman's view forward and may leave the radar prone to having its picture blotted out by bright sunlight falling on the screen. A good alternative is to hang the display from the wheelhouse deckhead – making sure that you're not going to hit your head on it every time you stand up!

Bear in mind, too, that most people find it much easier to orientate themselves to the radar if it's lined up with the boat's centre line and positioned so that the operator is facing forward. Second-best position is facing aft: athwartships-mounted radars are definitely confusing.

Finally, of course, you should think of the maintenance man. Neither you nor he is going to be particularly pleased if he has to take half the boat apart to get at the back of the set. Many modern radars lend themselves to being built-in or flush-mounted, but it's important to ensure that there is good ventilation to the rear of the set, reasonable access to the cable connections, and that the whole display can be taken out quickly and easily.

WIRING IT UP

The radar's power unit – which takes the nominal 12V or 24V of the boat's power supply and converts it to the different voltages required by the set's various components – is capable of dealing with quite significant voltage fluctuations. Like most electronic equipment, however, it much prefers a 'clean' supply, coming straight from the battery through its own fuse or circuit breaker.

The most difficult part of the wiring job is often routing the thick, multi-core cable linking the display unit and the scanner, whose size and stiffness preclude tight bends, while factory-fitted plugs make it impossible to thread it through small holes.

Some modern radars use thinner, more flexible cable, and of course plugs can, if necessary, be removed. A particular difficulty arises where the display-scanner cable has to go through the deck. On some boats, especially sailing boats whose masts may have to be lowered or removed, this can be achieved using waterproof deck plugs and sockets. This would be fine if only deck plugs were truly waterproof and could be relied upon to give a good connection. Personally, though, I don't like it: a much better solution is to run the cable through a well-sealed deck gland and either leave enough slack to cope with lowering the mast or accept the need to disconnect one end of the cable whenever the mast is unstepped. If the worst comes to the worst, you can always take the cable to a junction box inside the boat, making sure to leave a 'drip loop' to stop any water which makes it through the gland from getting into the box.

It is tempting to find one good route between the instrument panel and the mast, and use that for all the electronics. Don't! Radar and radio-telephone cables both carry radio frequencies at relatively high

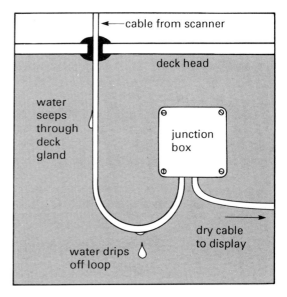

▲ Always leave a drip loop in any wiring that enters the boat from outside. Stray water drips off the bottom of the loop before it gets to the terminals.

power compared to the tiny voltages in the lines from receiving aerials. Manufacturers do their best to reduce the amount of power radiated from their cables, but even so, it is best to reduce the risk of mutual interference by keeping cable runs and aerials well separated.

TROUBLESHOOTING

Interference of any kind is usually fairly easy to identify: radar interference with a VHF RT, for instance, produces a high-pitched whine, while interference in the opposite direction appears as a series of bright radial lines on the radar screen whenever the transmit switch of the radio is pressed. One particularly puzzling example of radar interference is that it can make digital echo-sounders read zero! However it appears, though, interference stops when the offending equipment is switched off.

Curing interference once it has been identified is quite a different matter. Essentially it's a question of ensuring that the equipment concerned is properly earthed, that the power supply to the affected equipment is 'clean' and that cables are properly screened. Despite this apparent simplicity, curing interference often turns out to be extremely complicated, and quite beyond the scope of this book.

That, in fact, applies to most repair or maintenance work on radars: like most solid-state electronic equipment there is little the user can or should do apart from following the instructions and protecting the set from water, heat, and physical damage.

A sensible user's maintenance routine involves no more than periodically checking the physical security of any mounting bolts and exposed screws on the cover; keeping any external wiring connections clean and smeared with a thin film of petroleum jelly; and keeping the set itself presentably clean.

Radars are generally pretty reliable, so if yours doesn't work at all the first thing to check is that it is receiving power and that you have followed the correct start-up procedure. If the battery is switched on, the fuse or circuit breaker is OK, and all the connections along the supply line are clean and tight but the set is dead, try checking the internal fuse, usually located at the back of the display unit. Before changing it, though, make sure that there is nothing fouling the scanner, and that any external wiring is properly connected.

More complicated repairs are a job for a professional. Marine electronics specialists put a high price on their time, but you can console yourself with the thought that carrying out the repair itself is unlikely to take very long: it's the process of diagnosing the trouble that takes time.

A comment as simple as "It doesn't seem to have the range it had last season" is so much more useful than "It's not working properly" that it could save you a fortune. It's better still if you can not only tell the repair man exactly what the problem is, but whether it happens all the time or just occasionally, what seems to make it better or worse, and whether it happened gradually or suddenly.

12. Conclusion

In the first chapter of this book I commented that radar is fun to use. That's because it involves the operator in more direct interaction, interpretation, and control than any other navigation aid.

But getting the best out of it requires a degree of skill – and the only way to acquire that skill is by practice.

So please don't leave it until you're in thick fog or pitch darkness. Turn your radar on and play with it in daylight; get to know the controls of your particular set, and – most important of all – get used to interpreting the picture on the screen. Have fun!

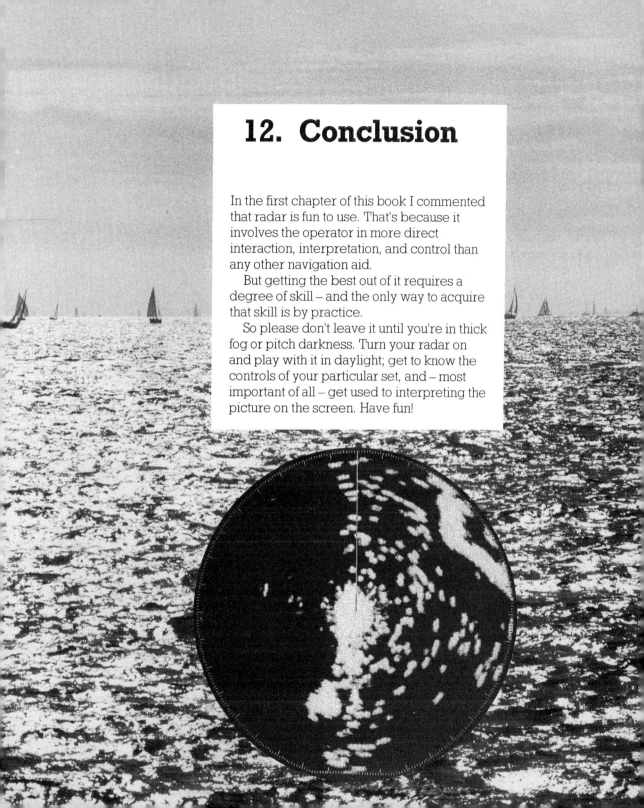

Glossary

Amplification Increasing the amplitude of a wave, as in making a sound louder, a light brighter, or increasing the power of a radio wave.

Analogue Used to describe the older style of radar display.

Beamwidth The width of a radar beam: more precisely, the angle over which the power of the radar signal is at least 50% of its maximum.

Blind arc An area shielded from radar transmissions by part of the ship's structure.

Brilliance A control which regulates the brightness of the radar picture.

Cathode ray tube The component which creates the radar picture.

Chinagraph A grease pencil.

Clearing line A line of position (cf) which is known to pass clear of a hazard.

Closest point of approach (CPA) The point at which an approaching vessel will be closest to your own.

Cocked hat The traingle formed by the intersection of three position lines.

Compass Refers to directions defined using a magnetic compass before correction for deviation and variation.

Compass safe distance The minimum distance from a compass at which a piece of electrical equipment can be installed without causing more than 1° deviation.

Contact The bright blob on a radar screen which represents the position of a target (cf).

Course The direction in which a vessel is intended to be moving through the water – not the same as its track (cf) or heading (cf).

Course-up Describes a radar display in which the picture is compass-stabilised so that the vessel's intended course is straight up the screen.

CPA Closest point of approach.

Cursor An electronically-generated block or cross used to indicate a position on a raster-scan display; or the rotating grid of index lines on an analogue display.

Decca The Decca Navigator System – a medium-range hyperbolic position-fixing system.

Deviation A compass error affecting magnetic compasses caused by the vessel's own magnetic field.

Differentiation An alternative name for the rain clutter circuit.

Discrimination A radar set's ability to show targets which are close to each other as separate contacts.

Drift The rate at which a boat moves due to the effect of a current or tidal stream.

EBL Electronic bearing line – an electronically-produced line radiating from the centre of the screen which can be moved around the screen by the operator to help in taking bearings.

Echo The returning radar signal reflected from a target (cf contact, target). Sometimes used as a synonym for contact.

Estimated position A vessel's calculated position, taking into account the effect of its speed, course, leeway, set, and drift.

Fix A vessel's known position.

Gain Amplification.

Guard zone An area of the radar screen defined by the operator, within which the presence of a contact causes an alarm to sound.

Head-up A radar display mode in which the vessel's heading is straight up the screen.

Heading The direction in which the vessel is pointing at any moment.

Heading mark A bright line on a radar screen representing the vessel's heading – sometimes called the heading flasher.

Horizontal beamwidth The beamwidth (cf) of a radar measured horizontally.

Index line A line drawn on the face of a radar screen or plotting sheet, usually parallel to the vessel's intended track, to assist in radar pilotage.

Interfacing Connecting two electronic navigation aids together so that information can be passed from one to the other.

Leeway A vessel's sideways movement through the water caused by the wind.

Liquid crystal diode (LCD) An electronic component whose colour can be made to change by stimulating it with a voltage. Banks of liquid crystal diodes can be connected together to form a display.

Loran Loran C – A long-range hyperbolic navigation system.

Magnetic Describes directions relative to magnetic north. Differs from True by an error known as variation, caused by the Earth's magnetic field.

Magnetron An electronic valve which uses a powerful permanent magnet to produce pulses of microwaves.

Microwaves Electromagnetic waves whose wavelength and frequency are between those of radio waves and infra-red.

NMEA The National Marine Electronics Association – an American organization responsible for defining most of the standard interfaces used in marine electronics.

Noise Random electromagnetic interference appearing on a radar screen as minute specks of light.

North-up A radar display mode in which the picture is compass-stabilized and rotated so that north is straight up the screen.

Open scanner A radar whose aerial is not enclosed within a radome (cf).

Patch aerial A radar aerial made up of a carefully-designed cluster of small copper pads.

Pill box A radar aerial made up of thin sheets of folded metal, focusing the radar beam in much the same way as the reflector of a torch.

Pixel A picture cell – one of several thousand tiny squares which together make up the picture on a raster-scan radar.

Plan position indicator (PPI) The most common type of radar display format, in which contacts appear in the form of a map or plan.

Plotting A system of observing and recording the movement of radar contacts.

Plotting sheet Paper forms, representing the radar screen, for plotting purposes.

Position line A line on which the vessel's position lies.

Pulse   A short transmission of radio waves or microwaves.

Pulse length   The duration of a pulse.

Pulse repetition frequency   The number of pulses per second.

Quantized   Describes a radar display in which each pixel can appear at any of several levels of brightness.

Racon   A radar transponder: a device fitted to some important navigation marks which transmits its own signal when stimulated by transmission from a nearby radar set. Causes a distinctive flash on the radar screen.

Radar reflector   Generally, any object that will reflect radar waves; specifically, a safety device intended to increase the strength of the echo produced by small craft or navigation marks.

Radarscope   A synonym for 'display' – common in the USA.

Radiator   A somewhat unusual synonym for aerial, antenna, or scanner.

Radome   A cover completely enclosing the rotating radar aerial and (usually) most of the transmitter and receiver circuitry.

Rain clutter   Weak contacts caused by radar echoes from rain, snow or hail.

Range   Distance – either the distance between an object and your own vessel, or the maximum distance at which the radar is set to operate.

Range rings   Rings on the radar screen representing several pre-determined ranges.

Raster-scan   A type of radar display whose picture is produced by digital analysis of the range and bearing of each contact, whose picture is made up of pixels.

Resolution   A synonym for discrimination.

S Band   Microwaves whose wavelength is in the order of 10cm – used for some big-ship radars.

Satnav   Any of several position fixing systems using radio transmissions from artificial satellites orbiting the Earth.

Scanner   A rotating aerial; or the unit of which the aerial is a part.

Sea clutter   Spurious contacts caused by radar echoes from waves.

Set   The direction of movement of a current or tidal stream.

Shadow areas   Areas shielded from radar transmissions by solid objects outside the vessel.

Sidelobes   Radio energy propagated by the aerial outside its beamwidth.

Slotted waveguide   A type of radar aerial made up of a length of precision-made metal tube with a series of accurately-milled slots in one side.

Standby   When the radar is switched on and most of its systems are functioning, but the transmitter is not transmitting.

Strobe   A range ring whose range can be varied by the operator (cf VRM).

Sweep   A rotating radial line in the picture produced by an analogue radar, and artificially simulated in some raster-scan sets.

Target   Any solid object within the operating range of a radar set.

Track   A vessel's movement relative to the ground.

Tracker ball   A ball, built into the control panel of a radar set, and used to control the movement of a cursor.

Transit   Two or more objects which lie on the same bearing from an observer i.e. which appear in line with each other.

True   Directions referred to true north, ie referred to a line joining the Earth's north and south poles. Differs from Magnetic by an error known as variation, caused by the Earth's magnetic field.

True motion   A radar display which takes account of the transmitting vessel's own movement, so as to show the movement of each target relative to the ground.

Vertical beamwidth   The beamwidth (cf) of a radar measured vertically.

VRM   Variable range marker – a range ring whose radius can be varied by the operator to help measure ranges accurately.

Waveguide   A hollow metal duct, used to conduct microwaves within a radar set.

X Band   Microwaves whose wavelength is in the order of 3cm – used for most marine radars.

**Fernhurst Books is the world's
leading nautical publisher.**

For a free full-colour brochure phone, fax or write to us:

Fernhurst Books, Duke's Path, High Street, Arundel,
West Sussex, BN18 9AJ.

Phone 01903 882277 Fax 01903 882715

Or visit our web site http://fernhurst.com

HEAD UP ☐
COURSE UP ☐   RANGE ....
NORTH UP ☐

CONTACT A

INITIAL RANGE ....
INITIAL BEARING ....
TIME OF 1ST PLOT ....

360
350   010
340   020
330   030
320   040
310   050
300   060
290   070
280   080
270 ------------------------ 090
260   100
250   110
240   120
230   130
220   140
210   150
200   160
190   170
180

CONTACT B

INITIAL RANGE ....
INITIAL BEARING ....
TIME OF 1ST PLOT ....

CONTACT C

INITIAL RANGE ....
INITIAL BEARING ....
TIME OF 1ST PLOT ....

This page only is copyright free